T0137062

Computational Social Sciences

Computational Social Sciences

A series of authored and edited monographs that utilize quantitative and computational methods to model, analyze and interpret large-scale social phenomena. Titles within the series contain methods and practices that test and develop theories of complex social processes through bottom-up modeling of social interactions. Of particular interest is the study of the co-evolution of modern communication technology and social behavior and norms, in connection with emerging issues such as trust, risk, security and privacy in novel socio-technical environments.

Computational Social Sciences is explicitly transdisciplinary: quantitative methods from fields such as dynamical systems, artificial intelligence, network theory, agent based modeling, and statistical mechanics are invoked and combined with state-of the-art mining and analysis of large data sets to help us understand social agents, their interactions on and offline, and the effect of these interactions at the macro level. Topics include, but are not limited to social networks and media, dynamics of opinions, cultures and conflicts, socio-technical co-evolution and social psychology. Computational Social Sciences will also publish monographs and selected edited contributions from specialized conferences and workshops specifically aimed at communicating new findings to a large transdisciplinary audience. A fundamental goal of the series is to provide a single forum within which commonalities and differences in the workings of this field may be discerned, hence leading to deeper insight and understanding.

More information about this series at http://www.springer.com/series/11784

Andrew Pilny • Marshall Scott Poole
Editors

Group Processes

Data-Driven Computational Approaches

 Springer

Editors
Andrew Pilny
University of Kentucky
Lexington, KY, USA

Marshall Scott Poole
University of Illinois
Urbana, IL, USA

ISSN 2509-9574   ISSN 2509-9582   (electronic)
Computational Social Sciences
ISBN 978-3-319-84053-6   ISBN 978-3-319-48941-4   (eBook)
DOI 10.1007/978-3-319-48941-4

Printed on acid-free paper

This Springer imprint is published by Springer Nature
The registered company is Springer International Publishing AG
The registered company address is: Gewerbestrasse 11, 6330 Cham, Switzerland

Contents

Chapter 1
Introduction

Andrew Pilny and Marshall Scott Poole

For many young group researchers, learning about advanced statistical methods can be quite the traumatic experience. Coupled with teaching, professional development, and being theoretical experts in their domain, fine graining the ins and outs of inferential statistics seemed like just another task on a full plate of work. Fortunately, for many of us, there was a rescuer. In 2000, Andy Field published his first book, *Discovering Statistics Using SPPS for Windows*, beginning a series of volumes dedicated to making statistics seem both easy and fun. Clarity was essential for Field, whose volumes always provided relevant examples (usually very humorous), clear screenshots, and example write-ups. Field's volumes were vital for not only learning about statistics, but reducing anxiety and uncertainty the complexities of inferential modeling.

However, the world has changed greatly since then, moving into what is generally referred to as the era of Big Data. Four characteristics generally characterize Big Data (Gandomi & Haider, 2015): (1) *volume* (i.e., bigger size and magnitude), (2) *variety* (i.e., more different types of data), (3) *velocity* (i.e., rate at which data is created), and (4) *complexity* (i.e., complex data structures that require cleaning and integration). But Big Data is not just about *data* per se, it is also about a new way thinking about measurement (King, 2016). For instance, instead surveying groups about their networks, we can now collect their interactions via their cell phones, email, and social media (i.e., trace data). Unfortunately, one of the consequences of Big Data is that many of the methods detailed by Field, which were exclusive variants of the general linear model, are either inappropriate or unsuited for much of the data we have on groups today. For instance, for data on online groups (e.g.,

A. Pilny (✉)
University of Kentucky, Lexington, KY, USA
e-mail: andy.pilny@uky.edu

M.S. Poole
University of Illinois, Urbana, IL, USA
e-mail: mspoole@illinois.edu

© Springer International Publishing AG 2017
A. Pilny, M.S. Poole (eds.), *Group Processes*, Computational Social Sciences,
DOI 10.1007/978-3-319-48941-4_1

coordination in Wikipedia), there can be millions of data points, which can results in nearly every independent variable tested being statistically significant. Likewise, interaction data from group members assumes a type of interdependence that violates many assumptions inherent in linear inference.

To address these issues, many researchers have called upon a paradigmatic change in thinking, largely referred to as *Computational Social Science* (CSS) (Cioffi-Revilla, 2013; Lazer et al., 2009). Computational social science is an interdisciplinary endeavor specifically tailored to handle the complexity of Big Data by merging together social science problems with computer science methods. As Wallach (2016) puts it, CSS can be thought of as research being undertaken by groups of "social minded computer scientists and computationally minded social scientists" (p. 317). The impact of CSS on group research has been especially notable. For instance, the new range of tools and thinking behind CSS has provoked innovative ways of understanding different group dynamics (e.g., Klug & Bagrow, 2016; Shaw & Hill, 2014) and collecting group data (e.g. Madan, Cebrian, Moturu, Farrahi, & Pentland, 2012; Radford et al., 2016).

Although the outlook of CSS is promising for the future of group research, there is a looming problem (e.g., Alvarez, 2016): for all the new work being produced using CSS methodology, there are few explicit avenues available to actually teach these methods. In other words, pedagogy is has taken a back seat to publication. The result is a sort of knowledge concentration or what boyd and Crawford (2012) refer to as a digital divide between the small minority who have access to Big Data and CSS resources and the majority who do not. Indeed, there are few graduate seminars, workshops (often expensive if they do exist), or handbooks that make it easy for the average social scientist to excel at CSS.

What is needed, therefore, is an "Andy Field book" for CSS, a resource to help *demystify* these methods and make it accessible to anyone willing to follow the white rabbit of CSS. To accomplish this goal, a resource would need to do several things. First, it would need to emphasize a *didactic*, rather than an inquiry-laden focus. That is, the primary objective is teaching rather than theory generation or original contribution to research. Second, it would need to be *transparent*, which means that codes and data should be shared and presented in a tutorial fashion. Transparency is vital in an age where we see social science continuing to be criticized for a lack of replication and secrecy regarding data and code. And finally, the resource should be *encouraging*. The spirit behind such an endeavor should reflect a growing notion that the more scholarly use of these methods, the better. As such, opaque and ambiguous language, equations, and procedures should be avoided in order to foster an environment that enables and empowers researchers to carry out a similar analysis.

These three values represent the spirit behind this book. The authors were given a relatively open format to write their chapters as long as it corresponded to a didactic, transparent, and clear avenue for anyone to pick up and take off with. The diversity of these chapters are quite evident: some are longer than others (e.g., Chap. 4: Relational Event Modeling), some introduce needed theoretical introductions (e.g., Chap. 6: Social Sequence Analysis), some use computer code (e.g., Chap. 2:

Response Surface Modeling), some use graphic interface programs (e.g., Chap. 5: Text Mining; Chap. 3: Bayesian networks) and some may not even use data at all (e.g., Chap. 8: Computational simulation).

Although no book on introducing CSS methods will be exhaustive, we aimed to provide the audience with what might interest group researchers the most. For instance, the growth of machine-learning is arguably one of the most dramatic changes in inferential modeling during the last twenty years (Hindman, 2015). Machine-learning models are often better equipped to handle Big Data because they are not dramatically influenced by sample size, often do not make crude normality assumptions, and have clear interpretations that explicitly acknowledge when the model predicts both accurately and inaccurately. As such, we included two chapters that explore different machine learning algorithms, Bayesian networks (Chap. 3) and decision-trees (Chap. 7).

Likewise, there has been a renewed increase in group dynamics that openly acknowledges time and order. In this case, group researchers can begin to seriously consider *dynamic* rather than *static* notions of emergence (Kozlowski, Chao, Grand, Braun, & Kuljanin, 2013). As such, Chap. 4 focuses on group interactions by viewing networks as relational events (i.e., episodic interactions), rather than relational states (i.e., enduring relationships). In this sense, relational event modeling can reveal dominant patterns of interactions by predicting ordered and even time-stamped histories of group interactions. Chapter 6 similarly focuses on time and order, but highlights social sequences of activities. One of the highlighted example of such a technique is that it can determine if group members behave in a synchronized pace (i.e., entrainment), provoking an important inquiry as to whether the emergence of group level properties are related to group performance.

It also important to recognize the new types of data that can be exploited by CSS methods. One example is the growing advent of analyzing text as data. In this sense, Chap. 5 explores text mining procedures and the development of semantic networks represented by co-occurrence relationships between different words and concepts. Sometimes there is not enough data or something was missing from data measurement. Chapter 9 deals with this through computational simulation with empirical data. Finally, sometimes we have enough data on groups with repeated observations that we can run quasi-field experiments. Chapter 2 adapts response surface methodology, a common method in the natural and physical sciences, to group research.

Lastly, as Alvarez (2016) notes, CSS is "developing at a dizzying pace" (p. 25). While researchers are rapidly developing tools to provide unique and sometimes ground-breaking insights into social inquiry, there is a need to pause and give back. Many of the tools used by CSS researchers were not developed individually in a vacuum. We owe a debt of gratitude to those who developed and taught us these methods, and owe it to the next and current generation of CSS researchers to share knowledge on how to use these methods. It can be seen as a sort of methodological "pay-it-forward". This book is one small attempt at such an endeavor.

References

Alvarez, R. M. (2016). *Computational social science: Discovery and prediction*. Cambridge, MA: Cambridge University Press.

Boyd, D., & Crawford, K. (2012). Critical questions for big data: Provocations for a cultural, technological, and scholarly phenomenon. *Information, communication & society, 15*(5), 662–679.

Cioffi-Revilla, C. (2013). *Introduction to computational social science: Principles and applications*. London: Springer.

Gandomi, A., & Haider, M. (2015). Beyond the hype: Big data concepts, methods, and analytics. *International Journal of Information Management, 35*(2), 137–144.

Hindman, M. (2015). Building better models prediction, replication, and machine learning in the social sciences. *The ANNALS of the American Academy of Political and Social Science, 659*(1), 48–62.

King, G. (2016). Preface: Big data is not about the data! In R. M. Alvarez (Ed.), *Computational social science: Discovery and prediction* (pp. vii–vi1). Cambridge: Cambridge University Press.

Klug, M., & Bagrow, J. P. (2016). Understanding the group dynamics and success of teams. *Open Science, 3*(4), 1–11.

Kozlowski, S. W., Chao, G. T., Grand, J. A., Braun, M. T., & Kuljanin, G. (2013). Advancing multilevel research design capturing the dynamics of emergence. *Organizational Research Methods, 16*(4), 581–615.

Lazer, D., Pentland, A. S., Adamic, L., Aral, S., Barabasi, A. L., Brewer, D., ... Gutmann, M. (2009). Life in the network: The coming age of computational social science. *Science, 323*(5915), 721.

Madan, A., Cebrian, M., Moturu, S., Farrahi, K., & Pentland, A. (2012). Sensing the "health state" of a community. *IEEE Pervasive Computing, 11*(4), 36–45.

Radford, J., Pilny, A., Reichelmann, A., Keegan, B., Foucault-Welles, B., Hoyde, J., et al. (2016). Volunteer science: An online laboratory for experiments in social psychology. *Social Psychology Quarterly, 79*(4), 376–396.

Shaw, A., & Hill, B. M. (2014). Laboratories of oligarchy? How the iron law extends to peer production. *Journal of Communication, 64*(2), 215–238.

Wallach, H. (2016). Computational social science: Towards a collaborative future. In R. M. Alvarez (Ed.), *Computational social science: Discovery and prediction*. (pp. 307–317). Cambridge University Press.

Chapter 2
Response Surface Models to Analyze Nonlinear Group Phenomena

Andrew Pilny and Amanda R. Slone

2.1 Introduction to Response Surface Methodology

Using Response Surface Methodology (RSM) is a lot like being a chef, mixing together different combinations of ingredients to see which ones come together to make the best dish. In this situation, strict linear thinking no longer applies. For instance, adding just the right amount of salt to a dish can bring out the sweetness in desserts or bump up the taste in more savory dishes. But, too much salt can overwhelm the flavor of a dish, just as too little salt can leave it tasting bland and unsatisfying. Chefs must find that perfect amount of salt that takes their dish from acceptable to exceptional. In addition, chefs must consider how the salt will interact with other ingredients in the dish. For example, salt interacts with the yeast in bread to help create texture, and it helps sausage and other processed meats come together by gelatinizing the proteins. Likewise, RSM helps us find the optimal amount of an outcome variable based on two or more independent variables.

This chapter will provide an introduction on how to use RSM to analyze nonlinear group phenomenon. First, the chapter will outline a brief history and background of the approach. Then, the chapter will walk the reader through a tutorial demonstrating how to execute the second-order model using the PROC RSREG function in SAS. Data previously collected from virtual groups in the game *EverQuestII* (see Williams, Contractor, Poole, Srivastava, & Cai, 2011) will be provided as an example.

A. Pilny (✉) • A.R. Slone
University of Kentucky, Lexington, KY, USA
e-mail: andy.pilny@uky.edu; amanda.slone@uky.edu

© Springer International Publishing AG 2017   5
A. Pilny, M.S. Poole (eds.), *Group Processes*, Computational Social Sciences,
DOI 10.1007/978-3-319-48941-4_2

2.2   Brief Background of RSM

Box and Wilson's (1951) treatment on polynomial models provided the foundation for RSM, which evolved and developed significantly (e.g., different variations of designs) during the 1970s (Khuri, 2006). Like many statistical methods, RSM developed in the natural sciences, but has yet to be applied extensively within the social sciences given the amount of repeated observations needed for RSM. Indeed, given the complexity involved in running controlled social science experiments and the typically low rate of manipulations, though, it is no wonder that RSM has not taken hold. However, with the advent of Big Data providing virtual Petri dishes of human behavior, RSM has garnered new interest in the social sciences for its ability to answer questions about complex group interactions (Williams et al. 2011. For example, due to its emphasis on optimization (i.e., finding the right combination of independent variables that maximizes a dependent variable), RSM has primarily impacted the world of business and performance management.

2.3   Basic Processes Underlying RSM

RSM is a blend between least squares regression modeling and optimization methods. More formally, RSM can be defined as the "collection of statistical and mathematical techniques useful for developing, improving, and optimizing processes" (Myers, Montgomery, & Anderson-Cook, 2009, p. 1). Moreover, instead of trying to only explain *variance*, RSM also seeks to clarify *optimization*. In other word, it is not necessarily about *how* a set of independent variables explains a dependent variable, but rather *what combination* of independent variables will yield the highest (or lowest) response in a dependent variable. In order to do this, RSM requires at least three variations in each variable, measured on a ratio or interval level.

To conduct an RSM test, there are typically five consecutive steps to go through (SAS Institute, 2013): (1) the regression modeling, (2) lack of fit, (3) coding of variables, (4) canonical analysis, and (5) ridge analysis. Each of these steps is described in more detail below.

2.3.1   Step 1: Second-Order Regression Modeling

The most common and most useful RSM design is the *second-order model* because it is flexible (i.e., not limited to linear trends), easy (i.e., simple to estimate using least-squares), and practical (i.e., has been proven to solve real world problems; Myers et al., 2009). The general linear model formula is identical to that which is used when conducting a regression (Eq. 2.1):

$$y = f\left(x_1, x_2\right) + e \tag{2.1}$$

In this general linear equation, y equals a response variable, x_1 and x_2 represent predictor variables, and e equals the error term.

But, RSM uses the second-order model in order to fully determine the *response shape* (i.e., the observed nonlinear trend). The equation for the second-order model is as follows (Eq. 2.2):

$$\hat{y} = b_0 + x'b + x'\hat{B}x \tag{2.2}$$

In this second-order matrix equation, "b_0, b, and $\hat{B}$ are the estimates of the intercept, linear, and second-order coefficients" (Myers et al., 2009, p. 223) respectively. One thing to note is that, unlike most other second-order regression models used when conducting group research, the results provided in this model are preliminary. That is, the results are used to determine linear, quadratic, and interactional relationships between the independent variables, not to identify the response shape.

2.3.2 Step 2: Lack of Fit

Lack of fit is how well predicted repeated observations match the observed data. In other words, lack of fit of the second-order model indicates that the predicted values of the data do not look like the observed values (see Montgomery, 2005, p. 421–422). For example, though salt (independent variable) may be shown to influence taste (dependent variable) in a second-order model (i.e., statistically significant), when we compare the predicted responses to actual taste ratings (e.g., feedback from customers), there are major discrepancies. This indicates a poorly fitting model.

When we have more than one observation on an independent variable, there are several things to look out for when calculating lack of fit. First, it is important to differentiate *pure error* from *lack of fit error*. Pure error is more common in regression modeling is determined by looking at the sum of squares (Eq. 2.3) variability between each repeated observation of the independent variables (y_{ij}) and the average value of the response variable ($\bar{y}_i$):

$$SS_{PE} \sum_{i=1}^{m} \sum_{j=1}^{n_i} \left(y_{ij} - \bar{y}_i\right)^2 \tag{2.3}$$

Lack of fit error is different because it uses a weighted version of y_{ij} and looks at the actual observed value of the dependent variable, not the average. The Equation (2.4) can be calculated by taking the sum of the difference of the average value of the response variable ($\bar{y}_i$) the fitted value of the response variable ($\hat{y}_i$), and weighting it by the number of observations at value of the independent variable (n_i):

$$SS_{LOF} \sum_{i=1}^{m} n_i \left(\bar{y}_i - \hat{y}_i \right)^2 \qquad (2.4)$$

From there, an F-test (Eq. 2.5) can be derived using mean squares (MS) from both equations to determine whether or not a quadratic model is even necessary to replace a reduced first-order model. For instance, if the lack of fit test is not significant for a first-order model, then there could be a reasonable argument that a second-order model is not event needed:

$$F_0 = \frac{MS_{LOF}}{MS_{PE}} \qquad (2.5)$$

Likewise, if the test is statistically significant for a second-order model, then by Occam's Razor (i.e., law of parsimony), we have evidence that a quadratic model might not be appropriate.

2.3.3   Step 3: Coding of Variables

Despite requiring variables to be measured at the interval or ratio level, RSM does not simply examine multiple sets of linear relationships. Instead, RSM conducts an experiment of sorts, and organizes variables into conditions to see which results in the optimal output. As such, to make it easier to conduct the canonical analysis (Step 4) and ridge analysis (Step 5), recoding values is a convenient way to examine the response shape at multiple values of the independent variables. As Lenth (2009) put it, "Using a coding method that makes all coded variables in the experiment vary over the same range is a way of giving each predictor an equal share in potentially determining the steepest-ascent path" (p. 3). In addition to simplifying the calculation, recoding the variables also produces results with respect to the original values of the independent variables. A common way to recode variables, as in the SAS package, is to do the following (Eq. 2.6):

$$\text{Coded value} = \frac{\text{Original value} - M}{S} \qquad (2.6)$$

whereas "M is the average of the highest and lowest values for the variable in the design and S is half their difference" (SAS Institutive, 2013, p. 7323). For instance, if there were five observations of on salt, ranging from two ounces to ten ounces, then the data for salt are stored in coded form using the following (Eq. 2.7):

$$x_{salt} = \frac{\text{Salt value} - (10-2)}{4} \qquad (2.7)$$

2.3.4 Step 4: Canonical Analysis of the Response System

The next step is to conduct a canonical analysis of each of the conditions. The purpose of the canonical analysis is to determine the overall shape of the data. For a first-order model, this is typically done through a method of *steepest ascent or descent*, wherein a linear shape determines which region of values creates an optimal response. However, for a second-order model, the shape can look more three-dimensional given the addition of interaction and polynomial terms. Here, we go back to our original Eq. (2.2) of a second-order response in matrix form (see Myers et al., 2009, p. 223):

$$\hat{y} = b_0 + x'b + x'\hat{B}x$$

To optimize the response ($\hat{y}$) and locate the stationary point (x_s) (i.e., the point of highest response in the dependent variable) we can set the derivative of $\hat{y}$ equal to 0:

$$\frac{\partial \hat{y}}{\partial x} = b + 2\hat{B}x = 0 \qquad (2.8)$$

and then solve for the stationary point:

$$x_s = -\frac{1}{2}\hat{B}^{-1}b$$

In these equations, *b* equals *a* vector of first-order beta coefficients:

$$\begin{bmatrix} \overset{s}{\hat{\beta}_1} \\ \cdot \\ \cdot \\ \cdot \\ \hat{\beta}_q \end{bmatrix} \qquad (2.9)$$

And $\hat{B}$ includes quadratic (diagonals) and interaction (off-diagonals) beta coefficients:

$$\begin{bmatrix} \hat{\beta}_{11,} & \hat{\beta}_{12}/2, & \cdots, & \hat{\beta}_{1q,}/2 \\ & \hat{\beta}_{22,} & \cdots, & \hat{\beta}_{2q,}/2 \\ & & \ddots & \\ sym. & & & \hat{\beta}_{qq,}/2 \end{bmatrix} \qquad (2.10)$$

For instance, consider if we trying to maximize taste ($\hat{y}$) with salt (x_1) pepper (x_2). After running a clean second-order model (i.e., no lack of fit), we find that:

$$\mathbf{b} = \begin{bmatrix} 3.65 \\ 4.69 \end{bmatrix}, \hat{\mathbf{B}} = \begin{bmatrix} -1.22 & -0.25 \\ -0.25 & -2.66 \end{bmatrix}$$

then

$$\mathbf{x}_s = -\frac{1}{2}\hat{\mathbf{B}}^{-1}\mathbf{b}$$

$$= -\frac{1}{2}\begin{bmatrix} -0.84 & 0.08 \\ 0.08 & -0.38 \end{bmatrix}\begin{bmatrix} 3.65 \\ 4.69 \end{bmatrix}$$

$$= \begin{bmatrix} 1.34 \\ 0.75 \end{bmatrix}$$

To compute this equation using matrix algebra, the following R code can be used:

```
B = matrix(c( 1.22,-.25,-.25,-2.66),nrow = 2,ncol = 2,byrow = TRUE)
b = matrix(c(3.65,4.69),nrow = 2,ncol = 1,byrow = TRUE)
x = -.5*((solve(B))%*%b)
x
```

As such, the predicted stationary point for taste based on salt and pepper (x_1, x_2) is 1.34 and 0.75. If the hypothetical fitted second-order model is

$$\hat{y} = 69.65 + 12.22x_1 + 3.45x_2 - 9.33x_1^2 - 6.32x_2^2 - 4.65x_1x_2$$

then the predicted highest response of taste $\left(\hat{y}_s\right)$ would be 63.63 by plugging in the optimal values for salt and pepper. It can then be re-expressed in the canonical second order form (this will be useful for later, see Montgomery, 2005, p. 446):

$$\hat{y} = \hat{y}_s + \lambda_1 w_1^2 + \lambda_2 w_2^2$$

$$= 63.63 + 1.34w_1^2 + 0.75w_2^2$$

where w_1 and w_2 are canonical variables (i.e., latent variables in relationship with the original independent variables).

From this point, it is necessary to determine the shape of the stationary point. The eigenvalues (λ) of the canonical analysis give indication to the nature of the shape (see Montogmery, 2005, p. 446):

$$| B - \lambda I | = 0 \tag{2.11}$$

If the eigenvalues for each independent variable are negative, then a *maximum stationary point* has been reached. A maximum stationary point looks like a hill, meaning that there is a point that indicates a high response. For most research, this is good news because it means that some combinations of variables entered in the model to produce a maximum response in the dependent variable. On the other hand, if they are all positive, then this indicates a *minimum stationary point*, meaning that the data will look like a *valley*. For most research, this is bad news because it means that some combinations of variables entered in the model to produce a minimum response in the dependent variable, unless a decrease in the dependent variable was what was desired of course.

Finally, if the eigenvalues are mixed, this indicates a *saddle point*, meaning that maximum or minimum solutions are not found, but rather multiple regions of high and low variables exist. In other words, the data will look like a series of hills and valleys, or perhaps even a plateau. For instance, a high value of x_1 and low value of x_2 may produce the highest value of y, while at the same time, a low value of x_1 and low value of x_2 may also produce the same value in y. Moreover, if they are all very close, or are at zero, then there is a flat area, meaning that there was little to no relationship between the independent variables and the response variable. Beyond looking at the eigenvalues, a two-dimensional contour plot is also a visual that can easily determine the shape of the response surface.

From our current example,

$$|B - \lambda I| = 0$$

$$\begin{bmatrix} -1.22 - \lambda & -0.25 \\ -0.25 & -2.66 - \lambda \end{bmatrix} = 0$$

By taking the determinant of the matrix:

$$(-1.22 - \lambda)(-2.66 - \lambda) - (-0.25 * -0.25) = 0$$

$$\lambda^2 + 3.88\lambda + 3.183 = 0$$

The solution, using basic completing the square calculus, is $\lambda_1 = -1.177$ and $\lambda_2 = -2.70$. As such, because both eigenvalues were negative, it indicates a maximum stationary point. This means that the canonical values for salt and pepper would yield the highest value of taste based on the data the researcher has collected.

2.3.5 Step 5: Conduct Ridge Analysis if Needed

Often when a saddle point is found, or if the researcher wants additional information regarding a maximum or minimum point, a ridge analysis can be performed. The purpose of a ridge analysis is to "anchor the stationary point inside the experimental region" and to give "some candidate locations for suggested improved operating conditions" (Myers et al., 2009, p. 236). In other words, the ridge analysis provides an estimated response value of y for each of the different values in the independent variables.

For instance, consider if the eigenvalue for pepper was essentially zero, but salt, as we found out, was significantly less than zero (see, Montgomery, 2005, p. 447). From this point, we would want to see what values of salt would yield a high amount of taste by analyzing the predicted response in taste from different values in salt. If the example formula was

$$\hat{y} = 69.65 - 14.87x_1 + 7.94x_2 - 0.33x_1^2 - 8.89x_2^2 + 13.65x_1x_2$$

and the resulting response in canonical form was

$$\hat{y} = \hat{y}_s - 13.56_1 w_1^2 - 0.02w_2^2$$

then we know we can pay more attention to salt because a single unit in the w_2 canonical variable would results in a 13.56 unit change rather than a small 0.02 unit change moved in the w_2 direction. In Table 2.1, a ridge analysis used this information to produce a line of predicted values that might indicate a trend:

Table 2.1 Example ridge analysis table

Order	Estimated response in taste	Un-coded value for salt	Un-coded value for pepper
1	65.24	2	1
2	59.17	3	0.9
3	49.53	4	0.8
4	36.32	5	0.7
5	19.55	6	0.6

From here, one can see how the decreasing levels of salt are related to a higher estimated response in taste, which could prove useful for future design of experiments.

2.4 RSM in Context

To demonstrate the usefulness of RSM in group research, this exemplar study employs data gathered from a download of data on 100,000 characters over 5 months in the Massive Multiplayer Online Game (MMOG) Everquest II (EQII).

2.4.1   About the Game

Commercially launched in November 2004, this game was estimated to have about 200,000 active subscribers in North America alone as of early 2008, the year in which the data was drawn from.[1] These players participate in thousands of teams over the 5 months, making it possible to draw much larger samples and making it possible to identify large samples of teams. Moreover, they incorporate precise metrics for the for team performance outcomes. As such, a random sample of 154 unique groups (i.e., no shared members) was analyzed for this tutorial.

As in most MMOGs, EQII players create a character and advance that character through challenges in a social milieu, typically banding together with other players for help and companionship. For each character, a class is chosen to fit some variation of the three basic archetypes found in nearly every fantasy MMO: damage-dealer, damage-taker and damage-healer. Each archetypal role has different capabilities, weaknesses and strengths, and the choice of class then determines how players develop their characters and how they will interact in the game environment and with other players. Players can communicate with others in the game through text messaging and voice chat.

Following a loose storyline, players use their characters to complete various tasks (quests) in order to earn virtual items such as currency and equipment. One important performance metric is number of "experience points" gained during a quest. Players must accumulate experience points to advance their character level. The character level is a fundamental indicator of players' success in the game. It not only represents a quantitative measure of players' skill and competence, but also determines whether players have access to certain quests and other game content, locations, and equipment. Until they attain the maximum level of 70, the accumulation of experience points is the only way for players to increase their character level. The amount of experience points associated with a given quest is associated with the difficulty of the quests and the value of the items won. Therefore, experience points can be used as a simple yet powerful indicator of players' performance at the common tasks in the game.

At the opposite end of the spectrum, a player can die during a quest. When a player dies in the game, they are not gone forever, but do pay a cost. For instance, for several minutes, the character is very vulnerable and cannot use many of their capabilities until they have had time to refresh many of their spells, buffs, and item effects. Moreover, their armor takes a significant amount of damage and if completely destroyed, the character will have to find a shop to get new armor or get it repaired. Finally, unless they are revived from a teammate, they will likely revive at a location far away from where the quest was being performed. As such, it is in the team's interest to avoid death because it can hinder their progress in the quest.

---

[1] There is no definitive evidence for the exact size of the population on Everquest II. The number 200,00 is estimated from multiple professional and fan sites such as http://www.mmogchart.com and http://gamespot.com

This study focused on group (i.e., heroic) quests and teams of three to six members. Generally, groups undertaking heroic quests include characters with different capabilities and skills. As discussed earlier, experience level is an important indicator of players' capabilities and competence, and groups often have members with different experience levels. The diversity in experience levels in a group can influence team processes substantially (Valenti & Rockett, 2008). Groups also typically are composed of members with different archetypal roles (i.e., damage-dealers, damage-takers, and damage-healers).

The groups in EQII that undertake heroic quests resemble the action teams described by Sundstrom, De Meuse, and Futrell (1990) in that they have short-term projects with clear goals and standards for evaluation, and members take on specific highly-interdependent roles. Their projects are the quests in the game, which require players to complete certain activities, such as finding objects or information, or killing a monster. Success or failure is clearly indicated by whether the quest is completed or not and whether or not members are killed during the quest. Analogous real world teams include military units, emergency medical response teams, and surgical teams.

2.5   Dependent Variable

2.5.1   Team Performance

Team performance was measured using two metrics. The first was the amount of *experience points* each player earned during the quest. These were obtained through the back-end database. Throughout the quest, characters earn points for successfully completing required tasks (i.e., defeating a monster, finding hidden objects). Likewise, *death* was the second and separate indicator of team performance. The total amount of deaths was calculated and the lower the number of group deaths, the better the performance.

2.6   Independent Variables

2.6.1   Complexity

Task complexity scores for each group were obtained through individually coding each quest. Detailed descriptions of each quest were obtained through *ZAM EverQuest II*, the largest EQII online information database. *ZAM* also features EQII wikis, strategy guides, forums, and chat rooms. Graduate and undergraduate researchers independently coded each quest based on the general definition of task complexity given by Wood (1986). According to Wood (1986), complexity entails three aspects: (1) component complexity (i.e., the number of acts and information cues in the quests), (2) coordinative complexity (i.e., the type and number of relationships among acts and cues), and (3) dynamic complexity (i.e., the changes in acts and cues, and the relationships among them). These features were used to code the complexity of each quest (*Mean* = 21.11, *SD* = 17.35, *Min* = 4, *Max* = 81).

2.6.2 Difficulty

Difficulty scores for each quest were obtained through Sony Online Entertainment. Each quest is given a static difficulty score ranging from 1 (least difficult) to 70 (most difficult). To create a variable that most closely resembled how difficult it was for the group attempting it, we subtracted the difficulty of the level of the quest from the highest player's character level. Thus, a negative number indicates that the group has at least one player that has a character level much higher that the quest they are attempting, meaning that it will likely be quite easy. On the other hand, a positive number indicates that everybody in the group has a character level below the quest difficulty level, meaning that it will likely be quite difficult to complete (*Mean* = –2.52, *SD* = 6.76, *Min* = –31, *Max* = 13).

2.7 Control Variables

2.7.1 Group Size

The more group members, the more likely there are opportunities for groups to both earn experience points and die. As such, to account for group size, we used group size as a covariate. Groups ranged from three (67.3 %), to four (22.2 %), five (5.2 %), and six (5.2 %) members. Since most groups has three members, the group size of three was used a reference point.

2.8 Data Analysis

The current example carries out RSM in SAS, through the proc. rsreg procedure. SAS is used here because it has perhaps the simplest code, though other programs can easily implement RSM like R and JMP.

2.8.1   Controlling for Group Size

Another benefit of using SAS is that the procedure, including contour plots and ridge analysis, are all done through specifying a few lines of code:

$$\text{data rsm;}$$
$$\text{set rsm;}$$
$$g6 = (\text{groupsize} = 6);$$
$$g5 = (\text{groupsize} = 5);$$
$$g4 = (\text{groupsize} = 4);$$
$$g3 = (\text{groupsize} = 3);$$

In the above line of code, the first thing that we must do is create the covariate variable. Since we want qualitative variable for each group size, we create four different variables and call them g6, g5, g4, and g3.

2.8.2   Experience Points: A Minimum Stationary Point

The next line of code runs the RSM procedure:

```
ods graphics;
proc rsreg data = rsm plots = (ridge surface);
model experience_pts = g6 g4 g5 g3 difficulty complexity / covar = 4 lackfit;
ridge max min;
run;
(odsgraphicsoff;)
```

The first line (ods graphics on;), simply tells SAS to turn on the ODS Statistical Graphics (Rodriguez, 2011). These graphics are necessary to produce the contour plots that show the predicted response based on different values of the independent variables. The second line of code does two things. First, it specified the data, which we have named "rsm" (**proc rsreg** data = rsm). Second, it tells the program which types of plots we want form the output. In this case, we want a ridge and surface plot (plots = (ridge surface)). The third line of code specifies the model variables. In model one, we are analyzing experience points as a function of quest difficulty and complexity while treating group size as a covariate. When reading this line of code, the dependent variable should come directly after the model term followed by an equal sign (model experience_pts=). The independent variables should come next (g6 g4 g5 g3 Difficulty Complexity), making sure to have the covariates come first. The covariate command lets the program know that the first four variables are to be treated as covariates and not included in the canonical and ridge analysis (covar = **4**).

The final line of the model command is the lack of fit test, telling the command to include it in the output (lackfit;). Finally, we want to include the ridge analysis to find values of the independent variables that predict a maximum or minimum response in experience points (ridge max min;). After these commands are properly arranged, we must tell the program to run it (**run**;). Turning the ODS Graphics off is useful because it might make future commands run a bit slower, even if they are not using the ODS Graphics.

2.9   Results

The following figures contain screenshots from the actual SAS output to ease in initial interpretation. Figure 2.1 contains the results from the least squares regression, including the interaction and polynomial terms. Before the results, however, are some descriptive information, including how the two independent variables were re-coded

The SAS System

The RSREG Procedure

Coding Coefficients for the Independent Variables		
Factor	Subtracted off	Divided by
difficulty	-9.000000	22.000000
complexity	42.500000	38.500000

Response Surface for Variable experience_pts: experience_pts	
Response Mean	3232.605656
Root MSE	1359.462094
R-Square	0.1533
Coefficient of Variation	42.0547

Regression	DF	Type I Sum of Squares	R-Square	F Value	Pr > F
Covariates	3	10950853	0.0348	1.98	0.1204
Linear	2	23611455	0.0751	6.39	0.0022
Quadratic	2	9855992	0.0314	2.67	0.0729
Crossproduct	1	3758993	0.0120	2.03	0.1560
Total Model	8	48177294	0.1533	3.26	0.0019

Fig. 2.1   Coding coefficients and ANOVA (experience points)

Residual	DF	Sum of Squares	Mean Square	F Value	Pr > F
Lack of Fit	120	235646022	1963717	1.55	0.1084
Pure Error	24	30485732	1270239		
Total Error	144	266131755	1848137		

Parameter	DF	Estimate	Standard Error	t Value	Pr > |t|	Parameter Estimate from Coded Data
Intercept	1	3661.765063	361.579354	10.13	<.0001	2326.341716
difficulty	1	-24.793149	29.353031	-0.84	0.3997	-587.744282
complexity	1	-53.178658	25.183288	-2.11	0.0364	-792.042760
difficulty*difficulty	1	3.050587	1.336332	2.28	0.0239	1476.484094
complexity*difficulty	1	1.246778	0.874219	1.43	0.1560	1056.020971
complexity*complexity	1	0.515613	0.342138	1.51	0.1340	764.267643
g6	1	1088.103762	514.943437	2.11	0.0363	1088.103762
g4	1	354.285863	276.326599	1.28	0.2019	354.285863
g5	1	292.070086	511.576337	0.57	0.5689	292.070086
g3	0	0	-	-	-	0

Factor	DF	Sum of Squares	Mean Square	F Value	Pr > F	Label
difficulty	3	13954485	4651495	2.52	0.0606	difficulty
complexity	3	27768937	9256312	5.01	0.0025	complexity

Fig. 2.2 Model coefficients and lack of fit (experience points)

for the canonical and ridge analysis, and descriptives for the dependent variable, which in this case is experience points ($M = 3232.61$). The omnibus analysis of variance table compares the different models (e.g., linear, quadratic, cross-product, covariate) to an intercept-only model in order to determine how much of an effect they add. For instance, because the quadratic terms by themselves ($F = 2.67, p = 0.07$), more or less, provide a better fit than an intercept-only model, it means they will likely be influential predicting an optimal or minimal response surface.

Figure 2.2 displays information on the lack of fit test and individual estimates for each independent variable. Overall, the lack of fit test was just above a 0.10 threshold for significance ($p = 0.11$). While this is generally acceptable as a rule of thumb, it points to some concern about how well the model predicted the actual response of experience points. Nevertheless, there were both linear and nonlinear effects in the model. For instance, there was a negative linear relationship with complexity ($t = -2.11, p = 0.03$), meaning that groups earned more experience points with less complex tasks. On the other hand, while there was not a linear relationship with difficulty ($t = -0.84, p = 0.40$), there was a quadratic effect ($t = 2.28, p = 0.02$),

Fig. 2.3 Canonical
analysis (experience
points)

The SAS System

The RSREG Procedure
Canonical Analysis of Response Surface Based on Coded Data

		Critical Value	
Factor	Coded	Uncoded	Label
difficulty	0.018236	-8.598818	difficulty
complexity	0.505573	61.964548	complexity

Predicted value at stationary point: 2271.661362

	Eigenvectors	
Eigenvalues	difficulty	complexity
1757.249593	0.882936	0.469494
483.502144	-0.469494	0.882936

Stationary point is a minimum.

meaning that there is a certain difficulty peak where groups tend to earn more experience points. To further investigate that point, a canonical analysis is useful here.

Figure 2.3 shows the results of the canonical analysis. Because the eigenvalues for difficulty ($\lambda_1 = 1757.25$) and complexity ($\lambda_2 = 483.50$) were both positive, this means the unique solution is a minimum. In other words, a unique combination of difficulty and class can yield a solution in which groups earned the *least* amount of experience points. As such, these two variables cannot tell us much about high performing groups, but do tell us a lot about *low* performing groups. Moreover, because difficulty was over three times the value of complexity, it means that experience points changes more rapidly along changes in difficulty compared to complexity. Finally, the table below gives the solution for the predicted minimum stationary point of 2271.66 experience points at a value of −8.59 for difficulty and 61.96 for complexity. Because the mean values for each variable is −2.52 and 21.11, this means that groups perform the worst when they choose quests that are about 40 units higher in complexity than average and when the groups highest member is about 8 units less than the quest value, which is higher than average.

Figure 2.4 shows the ridge analysis for a minimum solution. The quadratic effect for difficulty is clearly evidence here as the values fluctuate from moving higher from −9 to −7.87, then decreasing from −7.87 to −13.07. This is important because the relationship as demonstrated by the regression model is not linear, suggesting that the difficulty of the quest compared to the highest-level character in the group has a tipping point (∼−8.59). On the other hand, though there is an overall negative linear trend with complexity according to the regression model, the ridge solution paints a more complicated picture. For instance, almost equal predicted responses are obtained with a complexity value of 42.50 and 80.33. These results are in line with

The SAS System

The RSREG Procedure

Estimated Ridge of Minimum Response for Variable experience_pts: experience_pts				
		Standard	Uncoded Factor Values	
Coded Radius	Estimated Response	Error	difficulty	complexity
0.0	2477.237861	305.414405	-9.000000	42.500000
0.1	2393.253525	298.196900	-7.903294	45.837519
0.2	2335.568630	295.048816	-7.364132	49.648045
0.3	2298.871725	300.511618	-7.412100	53.710737
0.4	2278.533530	318.542869	-7.869005	57.772282
0.5	2271.681982	351.784851	-8.553852	61.734160
0.6	2276.769309	401.333098	-9.361888	65.591317
0.7	2292.959761	466.932201	-10.239946	69.362502
0.8	2319.777660	547.615192	-11.160508	73.067055
0.9	2356.935769	642.273088	-12.108464	76.720329
1.0	2404.251190	749.935315	-13.074990	80.333788

Fig. 2.4 Ridge analysis of minimum response (experience points)

the critical value threshold demonstrating that a complexity value near 60 is where groups are predicted to perform the least, much higher than average ($M_{compelxity} = 21.11$).

Although no maximum solution was found, a ridge analysis for maximum ascent tends to demonstrate simple linear effects for both difficulty and complexity (see Fig. 2.5). More specifically, groups are predicted to perform better when the character level of the highest character approaches the same level of the quest and the complexity of the quest increases. This makes sense because it means that the quest should not be as challenging for the group if they have at least one character in the group that is close to the quest difficulty level. The quest is still complex enough for group members to do activities that will give them a chance to earn points. However, no solid conclusions should be drawn from this. Instead, it may serve as an impetus to collect more data for future analysis.

Finally, Fig. 2.6 is a visualization of the response surface analysis as a contour plot, with covariates fixed at their average values. This means that this plot is most relevant for groups of three, which were the majority of groups playing this game. The minimum solution can be easily visualized by looking at the large ring representing values below 3000. Values closer to the center of that ring are the lowest predicted values of experience points. If you cross the intersection between the two critical values of −8.59 for difficulty and 61.96 for complexity, one can pinpoint to the center of the ring. The circles represent the predicted values for each observation.

The SAS System

The RSREG Procedure

Estimated Ridge of Maximum Response for Variable experience_pts: experience_pts				
		Standard	Uncoded Factor Values	
Coded Radius	Estimated Response	Error	difficulty	complexity
0.0	2477.237861	305.414405	-9.000000	42.500000
0.1	2591.514039	313.627581	-10.456265	39.614201
0.2	2738.304958	322.498612	-12.106776	37.047424
0.3	2918.733427	334.045714	-13.861558	34.688401
0.4	3133.381678	351.890047	-15.675575	32.465815
0.5	3382.571431	380.356503	-17.525517	30.335855
0.6	3666.492302	423.364871	-19.398557	28.271090
0.7	3985.262567	483.522055	-21.287199	26.253747
0.8	4338.959535	561.909924	-23.186815	24.271897
0.9	4727.635805	658.499613	-25.094415	22.317271
1.0	5151.328370	772.719727	-27.007990	20.383968

Fig. 2.5 Ridge analysis maximum response (experience points)

2.9.1   Model for Deaths: A Saddle Point

For the model with deaths as the response surface, we use the same code except switch the dependent variable form experience points to deaths:

```
ods graphics on;
proc rsreg data = rsm plots = (ridge surface);
model deaths = g6 g4 g5 g3 difficulty complexity / covar = 4 lackfit;
ridge max min;
run;
ods graphics off;
```

The initial outputs in Fig. 2.7 details similar information about the coded variables and analysis of variance.

As you can see in Fig. 2.7, there is a significant difference between an intercept only model and the linear, quadratic and cross-product models, suggesting that the

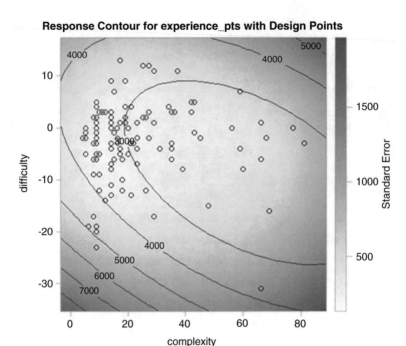

Fig. 2.6 Contour plot (experience points)

variables have considerable influence on deaths. However, as demonstrated by Fig. 2.8, the full model has a significant lack of fit, meaning that the average values of death deviate more than we would expect by chance from the predicted responses of deaths.

Indeed, although there are significant effects regarding the difficulty term ($t = 3.45$, $p < .01$) and overall interaction term (i.e., difficulty*complexity, $t = 2.38$, $p = .02$), the lack of fit finding puts a hitch into the entire analysis because it means that we cannot generalize much of the subsequent canonical and ridge analysis. From here, this usually means the researcher might look into some additional reasons for the lack of fit. For instance, there may not be enough variability in deaths and it might be useful to transform it to make it look more normally distributed (e.g., log linear transformation). Alternatively, the researchers might attempt to add more data or additional explanatory variables. Nevertheless, for demonstration, we will carry on with the canonical and ridge analysis.

As expected, there was no unique solution because of the saddle point response shape as demonstrated by the mixed signs of the eigenvalues (see Fig. 2.9). Nevertheless, the eigenvalue for difficulty ($\lambda_1 = 12.51$) is quite larger for complexity ($\lambda_2 = -1.03$), suggesting that there was more variability regarding changes in difficulty. Because there was a significant quadratic interaction, it is useful to look

The SAS System

The RSREG Procedure

Coding Coefficients for the Independent Variables		
Factor	Subtracted off	Divided by
difficulty	-9.000000	22.000000
complexity	42.500000	38.500000

Response Surface for Variable deaths: deaths	
Response Mean	3.718954
Root MSE	6.680704
R.Square	0.1725
Coefficient of Variation	179.6393

Regression	DF	Type I Sum of Squares	R-Square	F Value	Pr > F
Covariates	3	322.150612	0.0415	2.41	0.0698
Linear	2	394.317475	0.0508	4.42	0.0137
Quadratic	2	371.536805	0.0478	4.16	0.0175
Crossproduct	1	251.930606	0.0324	5.64	0.0188
Total Model	8	1339.935498	0.1725	3.75	0.0005

Fig. 2.7 Coding coefficients and ANOVA (deaths)

at a maximum ridge analysis to see what exact levels of difficulty were more associated with more deaths.

The ridge analysis complicates things even further because although the regression model suggested a nonlinear effect on difficulty, the ridge analysis does suggest a linear relationship (see Fig. 2.10). In other words, the more groups attempt quests that have difficulty levels higher than their highest level character, they are more likely to die in that attempt. Again, however, this might be due to a lack of fit.

Finally, the contour plot in Fig. 2.11 visually demonstrates the relationship between difficulty and complexity as it relates to the number of deaths incurred on a question. The wide open space in the middle indicates the least amount of deaths, but does not reveal a solution because those groups varied too widely on complexity and difficulty. Moreover, the bottom left and top right corners specify very high predicted values of deaths, meaning that no maximum solution could be found either because the existence of these high values occurs at seemingly opposite ends of the spectrum. That is, a high number of deaths can occur at a combination of either high complexity and low difficulty, or high difficulty and low complexity.

Residual	DF	Sum of Squares	Mean Square	F Value	Pr > F
Lack of Fit	120	6147.812868	51.231774	4.40	<.0001
Pure Error	24	279.166667	11.631944		
Total Error	144	6426.979535	44.631802		

Parameter	DF	Estimate	Standard Error	t Value	Pr > |t|	Parameter Estimate from Coded Data
Intercept	1	1.085448	1.776883	0.61	0.5422	1.055083
difficulty	1	0.128346	0.144247	0.89	0.3751	3.402473
complexity	1	0.060029	0.123756	0.49	0.6284	-0.058813
difficulty*difficulty	1	0.022638	0.006567	3.45	0.0007	10.956709
complexity*difficulty	1	0.010207	0.004296	2.38	0.0188	8.645243
complexity*complexity	1	0.000357	0.001681	0.21	0.8324	0.528477
g6	1	0.142864	2.530548	0.06	0.9551	0.142864
g4	1	1.764094	1.357931	1.30	0.1960	1.764094
g5	1	6.893767	2.514002	2.74	0.0069	6.893767
g3	0	0		.	.	0

Factor	DF	Sum of Squares	Mean Square	F Value	Pr > F	Label
difficulty	3	799.986242	266.662081	5.97	0.0007	difficulty
complexity	3	343.854037	114.618012	2.57	0.0568	complexity

Fig. 2.8 Model coefficients and lack of fit (deaths)

2.10 Conclusion

With the advent of mass amounts of data (e.g., trace data), it is possible to extract a bulk amount of information on how groups face different environments, process information, and perform. RSM inherently requires multiple observations on similar values of variables and is in a unique position to exploit such data. The main contribution of RSM is optimization. That is, through enough data collection, RSM can specify the conditions that are most likely to lead towards a certain outcome.

For instance, in the current example, traditional methods like regression and ANOVA would have been able to detect nonlinear relationships between difficulty and complexity, but they would not have been able to detect the specific values that can yield a certain outcome. The canonical analysis that RSM provides is more practical because it adds specific values and a contour plot that demonstrates how an outcome fluctuates based on different values of the independent variables. In this sense, the contour plot is a lot like a road map, guiding the researcher towards optimal paths that can yield insightful suggestions for practical implications.

Fig. 2.9 Canonical
analysis (deaths)

The SAS System

The RSREG Procedure
Canonical Analysis of Response Surface Based on Coded Data

	Critical Value		
Factor	Coded	Uncoded	Label
difficulty	0.079581	-7.249208	difficulty
complexity	-0.595284	19.581569	complexity
Predicted value at stationary point: 1.967925			

		Eigenvectors	
Eigenvalues	difficulty	complexity
12.515485	0.940705	0.339227
-1.030299	-0.339227	0.940705
Stationary point is a saddle point.		

The SAS System

The RSREG Procedure

Estimated Ridge of Maximum Response for Variable deaths: deaths				
Coded Radius	Estimated Response	Standard Error	Uncoded Factor Values	
			difficulty	complexity
0.0	1.815033	1.500875	-9.000000	42.500000
0.1	2.270613	1.406639	-6.822034	43.043526
0.2	2.968788	1.320360	-4.692571	44.071160
0.3	3.915048	1.248370	-2.587764	45.235399
0.4	5.110684	1.214818	-0.495253	46.455653
0.5	6.556150	1.256220	1.590387	47.704213
0.6	8.251648	1.405920	3.671825	48.969051
0.7	10.197280	1.676481	5.750511	50.244105
0.8	12.393103	2.060827	7.827300	51.525993
0.9	14.839153	2.545683	9.902726	52.812677
1.0	17.535451	3.119825	11.977140	54.102855

Fig. 2.10 Ridge analysis for maximum response (deaths)

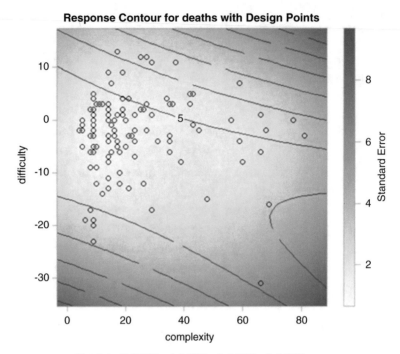

Fixed at: g6=0.0523, g4=0.2222, g5=0.0523, g3=0.6732

Fig. 2.11 Contour plot (deaths)

For instance, in EQII, groups are faced with decisions on which quests to attempt. Although traditional methods can detect relationships, they do provide an easy go-to guide that can be useful for actual decision-making. RSM, on the other hand, provides a very useful heuristic to help groups make decisions. For example, before a group attempts a quest, they can locate the values of the current quest and group (e.g., its difficulty) and pinpoint via the contour plot where their performance is predicted to land. If it lands on a very low performance spectrum, then this could be used as an important piece of information on whether or not that group should attempt to take on the quest.

Theoretically, RSM has the ability to test and examine a number of theoretical perspective. Notably, however, RSM has a unique opportunity to examine the basic tenets of chaos theory (see Tutzauer, 1996, for an application to organizations and groups), which highlights notions of unpredictability and unstableness. For instance, canonical and ridge analysis might not be very clean at times. That is, results that yield saddle points do not necessarily mean null findings. Instead, they have the ability to show how even small fluctions in the independent variables could cause dramatic changes in an outcome variable. Indeed, chaos theory would predict that in many contexts, a simple unique solution is not possible.

References

Box, G. E. P., & Wilson, K. B. (1951). On the experimental attainment of optimum conditions. *Journal of the Royal Statistical Society, B13*(1), 1–45.

Khuri, A. I. (2006). *Response surface methodology and related topics.* Singapore: World Scientific Publishing.

Lenth, R. V. (2009). Response-surface methods in R, using rsm. *Journal of Statistical Software, 32*(7), 1–17.

Montgomery, D. (2005). *Design and analysis of experiments.* Hoboken, NJ: Willey.

Myers, R. H., Montgomery, D. C., & Anderson-Cook, C. (2009). *Response surface methodology.* Hoboken, NJ: Wiley.

SAS Institute Inc. (2013). SAS/STAT 9.3 User's Guide. Cary, NC: SAS Institute Inc.

Sundstrom, E., De Meuse, K. P., & Futrell, D. (1990). Work teams: Applications and effectiveness. *American Psychologist, 45*(2), 120.

Rodriguez, R. N. (2011), An Overview of ODS Statistical Graphics in SAS 9.3, Technical report, SAS Institute Inc.

Tutzauer, F. (1996). Chaos and organization. In G. Barnett, & L. Thayler (Eds.), *Organization—Communication: The renaissance in systems thinking* (vol. 5, pp. 255–273). Greenwich, CN: Ablex Publishing.

Valenti, M. A., & Rockett, T. (2008). The effects of demographic differences on forming intra-group relationships. *Small Group Research, 39*(2), 179–202.

Williams, D., Contractor, N., Poole, M. S., Srivastava, J., & Cai, D. (2011). The Virtual Worlds Exploratorium: Using large-scale data and computational techniques for communication research. *Communication Methods and Measures, 5*(2), 163–180.

Wood, R. E. (1986). Task complexity: Definition of the construct. *Organizational Behavior and Human Decision Processes, 37*(1), 60–82.

Chapter 3
Causal Inference Using Bayesian Networks

Iftekhar Ahmed, Jeffrey Proulx, and Andrew Pilny

3.1 Introduction

The availability of new computational technologies, data collection opportunities, and data size is profoundly changing the nature social scientific analysis. Although traditional social scientific analysis (Content analysis, ANOVA, Regression, etc.) is still very much at the core of scholarly choice, newly found avenues are expanding analytical possibilities for social scientists. Prediction and network analyses are two of the areas impacted by newly found opportunities. Social scientists are now able to generate predictive results beyond traditional regression methods, thus are able to increase the power of social analysis. Hard sciences (i.e., Biology or Physics) have already developed a rich practice of collecting and analyzing massive amounts of data (Lazer et al., 2009). The possibility of dramatic changes in "analyzing, understanding, and addressing many major societal problems" became a reality due to an increase in the availability of informative social science data (King, 2011, p. 719). This data driven social scientific approach, popularly known as "computational social science", is a slowly growing field within social sciences largely spearheaded by interdisciplinary scientific teams (Lazer et al., 2009).

I. Ahmed (✉)
University of North Texas, Denton, TX, USA
e-mail: Iftekhar.Ahmed@unt.edu

J. Proulx
University of Illinois, Urbana, IL, USA
e-mail: proulx2@illinois.edu

A. Pilny
University of Kentucky, Lexington, KY, USA
e-mail: andy.pilny@uky.edu

© Springer International Publishing AG 2017
A. Pilny, M.S. Poole (eds.), *Group Processes*, Computational Social Sciences,
DOI 10.1007/978-3-319-48941-4_3

A number of new techniques utilized by present day computational social scientists are borrowed from computer science or information technology. Machine learning classification algorithm (MLCA) is one of such technique. MLCA is an umbrella term that consists of a variety of classification algorithms. The actual choice of a MLCA technique depends upon the theoretical and predictive interests of the researcher and the nature of data. Instead of looking for patterns in the dataset, MLCAs use cross-validation techniques to verify patterns in the data. MLCAs divide the data sample into several random samples, search for patterns in the earlier samples (except the last one), create probabilities or rules based on these patterns and then test those rules on the last sample. Bayesian network classifiers are one group of these MLCAs. Bayesian network MLCAs use *posterior probabilities (PP)* to generate classifications using Bayes' formula.

MLCAs became more user friendly for social scientists with the availability of Analytical Graphical User Interfaces (GUI). Weka is one of these available GUI for researchers. Developed at the University of Waikato, New Zealand, "Weka is a collection of machine learning algorithms for data mining tasks. The algorithms can either be applied directly to a dataset or called from your own Java code. Weka contains tools for data pre-processing, classification, regression, clustering, association rules, and visualization." (http://www.cs.waikato.ac.nz/ml/weka/). This chapter introduces Bayesian Network Analysis using WEKA.

"A Bayesian network consists of a graphical structure and a probabilistic description of the relationships among variables in a system. The graphical structure explicitly represents cause-and-effect assumptions that allow a complex causal chain linking actions to outcomes to be factored into an articulated series of conditional relationships" (Borsuk, Stow, & Reckhow., 2004, p. 219). Because of these links between actions and outcomes, social scientists can generate predictive results and develop network structure among variables beyond traditional social scientific approaches to increase the power of analysis. Conditional independence is at the core of Bayesian networks (Pe'er, 2005). Theoretically speaking, variable X is conditionally independent of variable Z given variable Y if the probability distribution of X conditioned on both Y and Z is the same as the probability distribution of X conditioned only on Y: $P(X|Y,Z) = P(X|Y)$. We represent this statement as $(X \perp Z|Y)$. Bayesian networks encode these conditional independencies with a graph structure (Pe'er, 2005, p. 1). A Bayesian network MLAs use *posterior probabilities (PP)* to generate classifications using Bayes' formula (Eq. 3.1):

$$P(y|x) = \frac{P(x|y) \cdot P(y)}{P(x|y) \cdot P(y) + P(x|not\ y) \cdot P(not\ y)}. \tag{3.1}$$

whereas $P(y \mid x)$ is a posterior probability of y (dependent variable) given x (independent variable), calculated by multiplying the likelihood of an attribute (x) given y ($P(x \mid y)$) and the class prior probability of y ($P(y)$) over that value time the probability of a false positive (P(x| not y) and the probability of a case not being y (P(not y)).

For instance, imagine you wanted to know if group performance (i.e., HIGH) is contingent on whether or not the group implemented a participative decision making (PDM) structure (i.e., TRUE). The Bayesian formula (Eq. 3.2) would try to determine $P(\text{HIGH} \mid \text{TRUE})$:

$$P\left(HIGH|TRUE\right) = \frac{P\left(TRUE|HIGH\right) \cdot P\left(HIGH\right)}{P\left(TRUE|HIGH\right) \cdot P\left(HIGH\right) + P\left(TRUE| \text{not } HIGH\right) \cdot P\left(\text{not } HIGH\right)} \quad (3.2)$$

For the current example, we will look at situations where a number of people are working together in a complex environment. A number of these situations like Military training or Firefighting are potentially dangerous and costly. The development of new technology such as games provides us an opportunity to train people in a safer environment. Because of technological development, we can make these training simulations very close to real world actions. We use the term multiteam system (MTS) to describe nested teams engaged in military or firefighting operations. MTSs are "teams of teams" where each team is nested within a larger collaborative group (Marks, DeChurch, Mathieu, Panzer, & Alonso, 2005). The purpose of the experiment was to investigate MTS collaboration dynamics in response to changes in the accuracy of the information environment surrounding teams.

Now consider a Military training simulation using games. Our experiment was conducted using a computer game called Virtual Battlespace 2 (VBS2) (see Pilny, Yahja, Poole, & Dobosh, 2014). VBS2 is a customizable combat simulation environment and is used globally for military training and simulation as it allows researchers to create custom scenarios where the researcher can add or remove stimuli in the simulation environment. Each experiment session lasted approximately three hours during which all participants engaged in two missions. All sessions where implemented in seven, sequential phases. For each mission, participants either played a scenario that contained entirely accurate information or a scenario that contained partially inaccurate information. In the experimental scenario, each MTS contained four participants divided into two teams of two people. Teams were tasked with navigating a map that contained landmarks and hazards along a route to the MTS's rendezvous point. As each team's location was unknown to the other team, therefore frequent communication was needed to coordinate activities.

This experiment uses participant's survey responses to see if we can predict which information condition the MTS assigned to groups. Each survey in this experiment was large and contained many scales and single response items, consequently providing a robust dataset. These are the types of data sets that we earlier mentioned as new possibilities for social sciences. As investigators, our interest is to identify factors that can predict information manipulation. However, the amount of data that we get and the research interest that we have together persuades us to explore new possibilities of social scientific research under the broader term "computational social science". Here, our particular interest is to see how Bayesian Network Analysis helps us in our investigation.

3.2 Scenario

This tutorial uses data from an experiment investigating MTSs. The purpose of the experiment was to investigate MTS collaboration dynamics in response to changes in the accuracy of the information environment surrounding teams. This chapter uses a subset of the original data. The following sections will walk the reader through the procedures used during data collection, variable selection, and the steps taken to prepare data for analysis.

Participants ($n = 129$) included undergraduate students from a large Midwestern university who were recruited through flyers and course announcements. A total of 38 MTS experimental sessions were conducted. Each MTS was comprised of four participants, divided into two teams of two. Of these 38 sessions, 33 yielded usable data as five sessions needed to be discarded due to recording errors.

Teams were tasked with navigating a map that contained landmarks and hazards along a route to the MTS's rendezvous point. As teams' location was unknown to each other in the MTS, field teams needed to use radio communication to coordinate a synchronous arrival at the rendezvous point. Once arriving at the rendezvous point, teams were given the task of eliminating a group of enemy insurgents. As a collective, the MTS was given three tasks: (1) to record landmarks the team navigated to for reconnaissance, (2) to successfully disarm and neutralize hazards such as explosive devices and insurgent ambushes, and (3) to coordinate a synchronous arrival at the rendezvous point.

While traveling to their rendezvous point, teams were exposed to pre-recorded radio messages that were intended to represent orders from the MTS's commanding officer. These radio messages took the form of audio played through each participant's headset that played when teams reached certain locations in the simulation. These messages gave teams information regarding the path that lie directly ahead of them and were also used to assign teams tasks such as confirming that an object exists along the path (e.g., a suspicious looking backpack) or exiting their vehicle to disarm an explosive device. Teams were exposed to ten messages during each mission. In the accurate condition all ten messages contained correct information that teams could verify within the simulation (e.g., if they are told there is a suspicious abandoned vehicle ahead, the suspicious vehicle actually existed). In the inaccurate condition two of the ten messages were inaccurate (e.g., if a team is told there is a suspected explosive device ahead, there was no explosive device).

All self-report items were measured at the individual level and observed measures were coded for each team in each session. As observed data were aggregated to the team level, there are two observations per session, one for each team. Screen recordings (videos) were used to construct a behaviorally anchored coding system and each MTS was coded for the five outcomes used in this analysis. Two independent raters coded each video and coding was largely objective. Kappa was used to measure interrater reliabilities and exceeded .90 in all cases thus suggesting an acceptable level of agreement.

In this tutorial, we will be using the participant's survey responses to predict manipulation of information condition in the MTS missions. Each survey in this experiment was large and contained many scales and single response items. Additionally, there are several observed outcome variables recorded for each mission. In this case, we chose 118 variables that are a mixture of scales, single survey items, and outcome variables to demonstrate how Bayesian networks can be used to accommodate robust datasets.

3.2.1 Variables

118 variables from the dataset were initially explored in this tutorial (1 dependent and 117 independent). We will be using *information accuracy* as our dependent variable (it is called MA_Accurate). Information accuracy was manipulated as a counter-balanced fixed effect in the experiment as each team was randomly assigned to either the accurate or inaccurate condition for their first mission. For the accurate information condition, all information given to teams was accurate, for the inaccurate condition, information given to teams contained two erroneous pieces of information. The remaining variables are treated as independent variables in this tutorial.

3.2.2 Data Preparation

Prior to analysis, data need to be cleaned and formatted. In this case, our data preparation involved two steps. First, all data were merged into a single file containing all of the variables we will use in our analysis. This means that survey responses, observed outcomes, and a dummy coded variable indicating which information condition the participant was in were combined into a single file. All variables are numeric.

After merging data, we removed any cases that containing missing values. In this case, a recording error occurred resulting in five sessions with partially mission data. These sessions were removed list wise. In order to analyze the dataset in Weka environment, we created a comma delimited file (.csv). MLCs work best with Binary Dependent Variable that we are going to predict. However, we can also use Nominal Variable with more than two categories.

3.3 Description of Weka Environment

This section describes the Weka GUI and how to explore different options to run an analysis. Figure 3.1 shows the opening window. The Explorer button allows us to locate and choose the data file that we are going to use. Once you click the explorer tab, it will open a window that provides an Open File option (Fig. 3.2). That option helps us to explore our data location and choose the file we will use for this experiment.

Fig. 3.1 WEKA main GUI

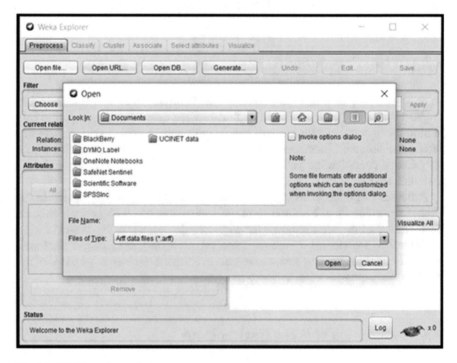

Fig. 3.2 WEKA explorer window

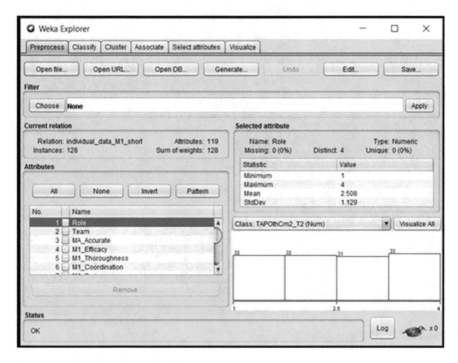

Fig. 3.3 Variables and basic statistics

Once the file in loaded in Weka, you can click any variable and see the basic statistics including maximum and minimum value, mean, standard deviation, and also visual representation of the data (Fig. 3.3).

Figure 3.4 shows the variable that we are going to predict. It shows that the variable is nominal with two categories. As you can see, category I has 63 cases and category A 65 cases. It is recommended to have either exact or very close number of cases in categories for better prediction results because classification problems can occur from imbalanced data. A data set is imbalanced if the classes are not represented equally within the data set. It is common to have data sets with small imbalances. However, large imbalances would definitely cause a problem. The best way to tackle the problem is by collecting new data. If that is not possible, then the option is to generate synthetic samples to balance the classes. This synthetic sample generation usually randomly sample attributes from minority class instances. If there are discrepancies, Weka allows to 'under-sample' or 'over-sample' a category. Over-sampling in Weka resamples datasets by applying the Synthetic Minority Oversampling TEchnique (SMOTE).

The Classify tab allows us to run classification algorithms. Figure 3.5 shows us a list of classifiers available based on the nature of our data. The first option here provides different Bayes classifiers. For this experiment, we are using BayesNet classifier.

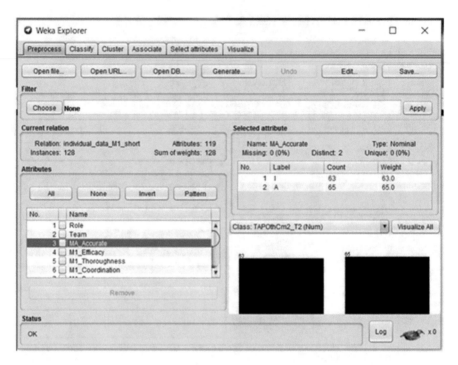

Fig. 3.4 Variables in prediction analysis

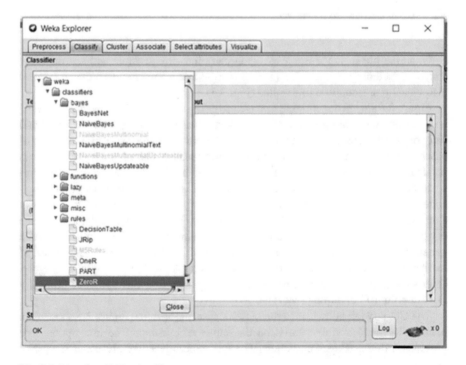

Fig. 3.5 List of available classifiers

3.4   Running Bayesian Network Analysis in Weka

3.4.1   Analysis with All Variables

There are 118 variables in our data set. MA_Accurate is the variable that provides us binary condition that we are interested to know. The rest of the variables will be used to predict MA_Accurate. There are three steps of running this experiment. First, in order to see the prediction power of our data set, we are going to run an analysis with all the variables. Having so many predictor variables is not a good practice as it makes explanation complicated. We are using it for two reasons. First to demonstrate advanced analysis possibilities so that we can use the technique if experimental situation demands such a robust analysis. Second, we like to compare results between all variables and few important variables that we are going to select later. Figure 3.6 shows us the basic window. As you can see, the classifier choice is BayesNet and the variable button shows MA_Accurate. It also shows that the variable is Nominal. To run the classification algorithm, we simply need to hit the start button. There is one additional step to remember. Figure 3.6 shows that, to determine the predictors of information accuracy, we are using a tenfold cross-validation method. It means that the algorithm will divide the sample into ten random samples. Then, it will use the first nine to create probabilities and search for patterns and

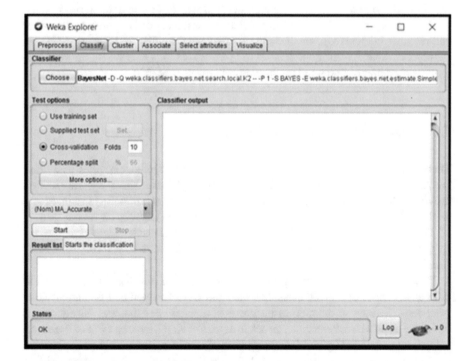

Fig. 3.6 Basic run window

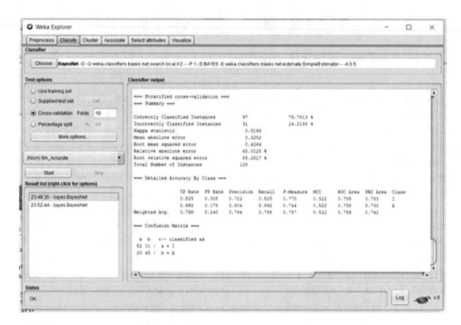

Fig. 3.7 Output window with run results

develop rules based on those patterns. Finally it will test those derived rules on the tenth sample. The number of cross validation choice depends upon the researcher and research interest (e.g., smaller samples may need smaller folds). A user can also supply a completely different data set to test the MLCA. Once we hit the start button, the classifier output window provides the result (Fig. 3.7).

3.4.2 Understanding Weka Output

There are three important sections of the output that together provides us a clear picture of our analysis. First is the *Stratified cross-validation Summary*. This section provides detail into the number of correctly and incorrectly classified instances and total number of instances. For us these were 90 (70.31 %), 38 (29.68 %) and 128.

The most important output for us is the second part of result - *Detailed Accuracy By Class*. Five important statistics for us are the *Precision, Recall, F-Measure, ROC Area*, and *Class* (Table 3.1).

First, the Weka output table provides the rate of true positives (TP Rate) or the ratio of instances of a given class that was correctly classified and the rate of false positives (FP Rate) or the ratio of instances of a given class that was falsely classified. Then it provides *Precision, Recall, F-Measure, ROC Area*, and *Class*. Precision is the ratio calculated by dividing proportion of true instances of a class by the total number of instances classified as that class. Recall is the ratio calculated by dividing

Table 3.1 Weka output for full model

	Precision	Recall	F-Measure	ROC area	Class
	0.667	0.794	0.725	0.770	I
	0.755	0.615	0.678	0.770	A
Weighted Avg.	0.711	0.703	0.701	0.770	

Table 3.2 Confusion matrix
for full model

a	b	← classified as
50	13	a = I
25	40	b = A

the proportion of instances classified as a given class by the actual total of that class. The F-Measure (Eq. 3.3) is calculated by combining precision and recall in the following manner:

$$F = 2 \cdot \frac{Precision \cdot Recall}{Precision + Recall}. \tag{3.3}$$

The overall F-Measure (mentioned in the output as Weighted Avg.) is the model accuracy. The accuracy of our test is shown by the ROC Area. A ROC Area value of 1 denotes a perfect test whereas a value of .5 is equal to random guessing. So, a worthwhile value is the one above .5 and better if that is closer to 1. In this test, our value is 0.701, which is good enough to accept results. Class denotes the values of our binary classes. Here 'I' represents *Inaccurate* and 'A' represents *Accurate* classes of our MA_Accurate variable. Based on our results, we can say that the test can identify whether the information scenario given by the researchers were accurate or inaccurate about seventy six percent of the time. Finally, the *Confusion Matrix* provides the statistics of how many times a particular class was classified rightly or wrongly. Our results indicate that in 50 cases of class I were classified as I (right classification) and in 13 cases as A (Table 3.2).

3.4.3 Assessing Information Gain

Although we have a good model that has a 70 % prediction power, a question of the power of individual variables in prediction remains. Although we have 118 variables in the test, it is good to find out how much each of these variables is contributing to the prediction analysis. *Assessing Information Gain* is one way that allows us exactly to do that. The reason behind this test is to identify and exclude variables that are not contributing much to prediction, eliminating them, thus make the model more parsimonious.

In order to run *Information Gain*, we need to go to *Select attributes* tab and choose *InfoGainAttributeEval* (Fig. 3.8). The select method will automatically

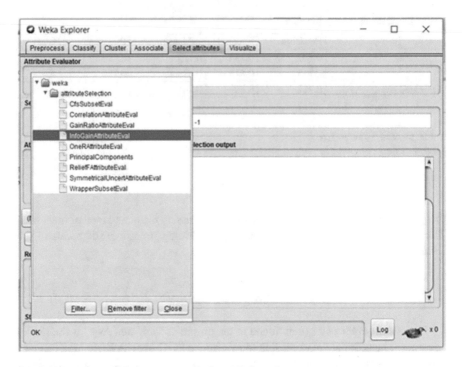

Fig. 3.8 Selection of *InfoGainAttributeEval in Select attributes* tab

change to ranker and will provide a window to accept that choice. Once we click OK it will be set. By clicking the start button, we will get the *Information Gain* results in the output area (Fig. 3.9).

In the *Ranked attributes* section, our results identified three variables with some information gain number (Fig. 3.10). The numbers are the amount of information gained from that particular variable (on the left side). Starting from variable '118 TAPOthCm2_T2' the numbers are all 0. It means that those variables are not contributing to any prediction analysis. Three variables our information gain test identified were team efficacy, team thoroughness, and speed. All variables were observed and measured at the interval level.

Team efficacy measures the degree to which teams accomplished their task of neutralizing hazards in the field. High scores of team efficacy were obtained by MTS's that identified and neutralized threats such as explosive devices and insurgent ambushes efficiently and quickly. Low scores of team efficacy indicate an MTS that did not neutralize threats, needed multiple attempts to eliminate threats, or took damage while completing a task. MTSs were placed into three categories based on their scores: (1) High, (2) Average, and (3) Low.

Team thoroughness measures the extent to which teams completed their task of recording the location of landmarks and hazards during their mission. High scores

Fig. 3.9 *Information Gain* results output

of team thoroughness indicate an MTS that correctly identified the name and location of mission landmarks and hazards. Low scores of team thoroughness indicate an MTS that did not accurately record the name or location of landmarks and hazards that they were tasked to locate. They were similarly placed into.

Speed was measured as the time in seconds that it took each team to complete the mission. Completion of the mission was denoted by the moment at which each team first arrived at the rendezvous point and was similarly placed into three categories based on one standard deviation: (1) Long, (2) Average, and (3) Short.

An analysis with only three identified variables with information gain statistics would yield almost similar result. As such, it is time for us to re-run the test with selected variables.

```
=== Run information ===

Evaluator:      weka.attributeSelection.InfoGainAttributeEval
Search:         weka.attributeSelection.Ranker -T -1.7976931348623157E308 -N -1
Relation:       individual_data_M1_short-weka.filters.unsupervised.attribute.Remove-R2
Instances:      128
Attributes:     118
                [list of attributes omitted]
Evaluation mode:    evaluate on all training data




=== Attribute Selection on all input data ===

Search Method:
        Attribute ranking.

Attribute Evaluator (supervised, Class (nominal): 2 MA_Accurate):
        Information Gain Ranking Filter

Ranked attributes:
 0.2561      3 M1_Efficacy
 0.1124      7 M1_Speed
 0.0901      4 M1_Thoroughness
 0         118 TAPOthCm2_T2
 0          41 TrstSq3_T2
 0          43 TrarSq5_T2
```

Fig. 3.10 *Ranked attributes* section of *Information Gain* result

3.4.4   Re-run with Selected Variables

In order to re-run the test, we need to go back to the Processes tab and select the variables that we need. Here we need M1_Efficacy, M1_Speed, and M1_Thoroughness. We also need our main variable MA_Accurate (Fig. 3.11). Once we select these four variables by clicking the checkbox beside them, we need to click the Invert button right above the list of variables area. This will reverse the selection and will select all variables other than the four we need. Now we can click the remove button right under the list area and remove all unnecessary variables from our analysis (Fig. 3.12). Once this selection process is done, we can replicate the analysis exactly as before.

Table 3.3 shows us our re-run results. As you can see, there is a slight decrease in the overall F-Measure from 0.701 to 0.656. However, the important part to know is that we have significantly decreased much of the noise in the data (i.e., variables that do not predict well), making the data much more interpretable and more substantially (rather than statistically) significant. A look at the probability distribution table can tell us more about the specific odds used to make prediction based on Bayes' theorem.

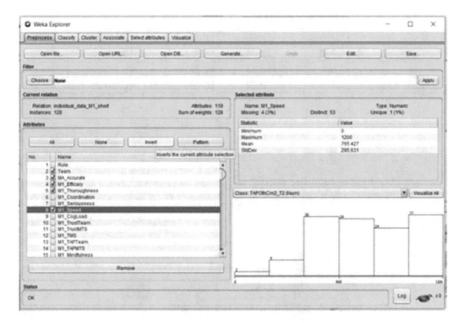

Fig. 3.11 Selection of variables with information gain

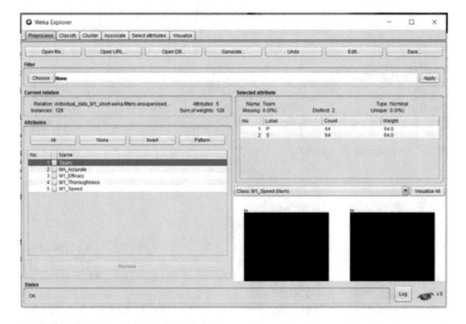

Fig. 3.12 Selected variables for final analysis

Table 3.3 Weka output for reduced model

	Precision	Recall	F-Measure	ROC Area	Class
	0.727	0.508	0.598	0.701	I
	0.631	0.815	0.711	0.701	A
Weighted Avg.	0.678	0.664	0.656	0.701	

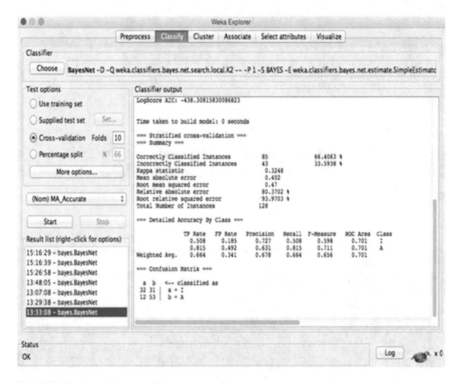

Fig. 3.13 Final prediction model

3.4.5 Probability Distribution

Weka allows viewing the graphical structure of the network. By right clicking the result in the explorer and a drop-down menu appears with a "Visualize graph" option (Fig. 3.13). The graph represents the graphical network with relationship among nodes (Fig. 3.14). This window allows inspecting both the network structure and probability tables. This graph is very useful to identify relationship between nodes. Each node in the graph represents a variable or condition and their relationships represent the network. It is similar to any other network structure.

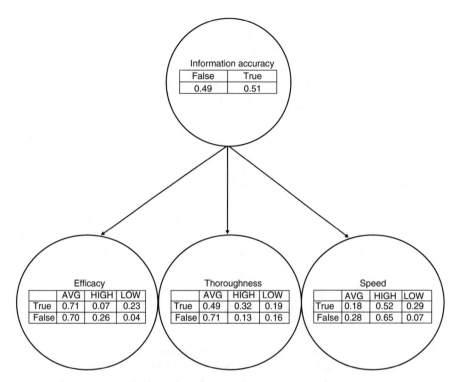

Fig. 3.14 Probability distribution table of one-parent model

Moreover, the network in the graph are directional indicating a directional relationship. If you place your cursor on any node, it will get high lighted. Clicking that node provides the probability table (Fig. 3.15). "The left side shows the parent attributes and lists the values of the parents, the right side shows the probability of the node clicked conditioned on the values of the parents listed on the left" (Bouckaert, 2004, p.29).

Using these probabilities, it is possible to calculate odds using Bayes' theorem. For instance, consider whether or not there was a relationship between the information manipulation and thoroughness (78 = AVG, 22 = HIGH, 28 = LOW). The corresponding probability of having a high thoroughness score and being in the true information accuracy group was 0.32. If we plug this into Bayes' theorem, we can determine the posterior probability of an MTS in the true information condition having a high score based on the probabilities given in Fig. 3.14 (see also Witten, Frank, & Hall, 2011, p. 260). To calculate theses, observe that Fig. 3.14 gives the odds of being in the True condition as 0.51 and 0.32 when thoroughness is high. The same odd when information accuracy is False is 0.49 and 0.13. To obtain conditional probabilities, we can use adapt Eqs. 3.1 into 3.4:

$$P(y|x) = \frac{P(x|y) \cdot P(y)}{P(x|y) \cdot P(y) + P(x|\text{not } y) \cdot P(\text{not } y)} \tag{3.1}$$

$$\frac{PP(\textit{True Info}) \cdot P(\textit{High Thorough}|\textit{True Info})}{PP(\textit{True Info}) \cdot P(\textit{High Thorough}|\textit{True Info}) + P(\textit{High Thorough}|\textit{False}) \cdot P(\textit{False})} \tag{3.4}$$

The values from Fig. 3.14 help us solve this equation:

$$\frac{0.32 \cdot 0.51}{(0.32 \cdot 0.51) \cdot (0.13 \cdot 0.49)} = 0.7193$$

As such, the conditional probability that an MTS had a high thoroughness score and was in the true information condition was 71.93 %, suggesting a significant relationship between having true information and better performance.

3.4.6   Re-run with Two Parent Nodes

Many social scientists and group researchers are interested in moderation or in other words, interaction effects. One way to get at this type of analysis is through increasing the parent nodes from one to two. This allows the predictor variables to interact with one another to create join probabilities. Indeed, one of the reasons it is called Naïve Bayes is because the predictor variables operate independent from one another.

To increase the amount of parent nodes from one to two, simply click on the BayesNet classifier next to the "Choose" button in Weka to open the generic object editor (Fig. 3.15). Then click on the "*searchAlgorithm*" box next to the "*Choose*" button and increase the "*madNrOfParents*" from one to two (Fig. 3.15). Finally, re-run the analysis (Table 3.4).

The table here is promising because the F-Measure has substantially increased from 0.656 to 0.702. Similarly, clicking on visualize graph will give us a probability distribution table (see Fig. 3.16). For instance, consider if we looked at efficacy and speed and wanted to determine is those groups who had high efficacy and average speeds:

$$\frac{0.51 \cdot 0.07 \cdot 0.838}{(0.51 \cdot 0.07 \cdot 0.838) + (0.49 \cdot 0.26 \cdot 0.091)} = 0.721$$

Here, we see that MTSs that had high efficacy and completed the mission in average times (i.e., not too long or short) had a 72.1 % chance of being in the true condition, demonstrating a significant relationship in how the manipulation may have influenced group performance.

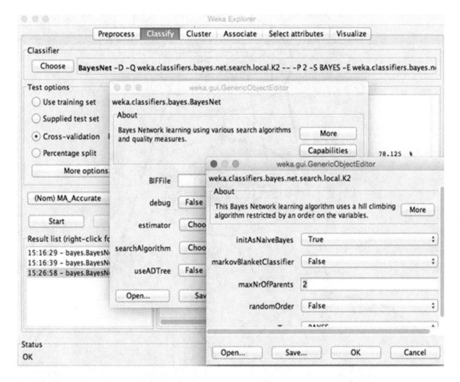

Fig. 3.15 Setting up a two-parent model

Table 3.4 Weka output for reduced model with two parent nodes

	Precision	Recall	F-Measure	ROC area	Class
	0.727	0.635	0.678	0.737	I
	0.685	0.769	0.725	0.737	A
Weighted Avg.	0.706	0.703	0.702	0.737	

3.5 Conclusion

This article demonstrated the opportunities offered by a data driven social scientific approach, popularly known as "computational social science". Here we explored a situation where a number of people were working together in a complex environment. These people constituted true groups as they were interdependent with common goal and fate. Receiving accurate information was vital in their success. However, information accuracy was manipulated to see the effect on group processes. It was a simulation of a real world group oriented problem, and due to recent technological developments, the simulation was very close to real world actions.

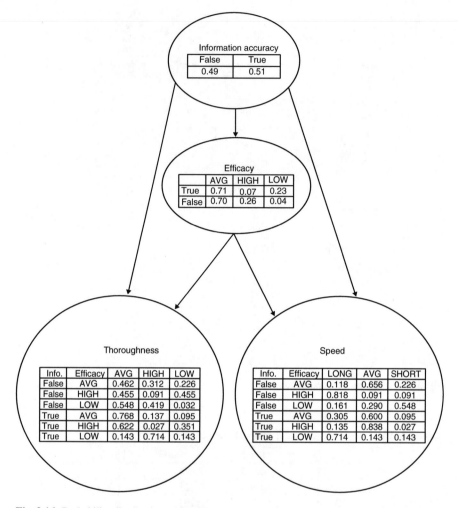

Fig. 3.16 Probability distribution table of two-parent model

Group communication scholars has been exploring and analyzing such situations for a long time. What made this situation unique is the number of variables that our system collected. We had 118 variables in the data set. What we observe here is the opportunity of collecting massive data. Previously, social scientists would limit the number of variables because of the complication that would arise in analysis and explanation. Data collection in those cases would be limited based on existing theories. Although theoretically sound, this line of research would be conservative in exploring many variables, limiting the possibility of discovering novel effects. Computational social science helps us to address this barrier.

Another possibility that comes forward is the opposite of theory driven analysis. Instead of an a-priory approach, now we can let the data show us relationships and relational patterns and make sense of the relationship later based on existing theories. During the process, this article demonstrates that the MLCA analysis could actually discriminate variables based on their importance in understanding the situation.

This article also demonstrates new ways of interpreting and presenting social scientific results. Here we not only see that the conditional probability that an MTS had a high thoroughness score and was in the true information condition suggesting a relationship between having true information and better performance, we knew that the probability was 71.93 %. Such accuracy derived from complex situations could be considered as a major improvement in social scientific analysis.

This demonstration represents one of many novel possibilities offered by computational social science methods to social scientific scholars. Together with traditional approaches, new methods would definitely enhance our explorations and analysis of social situations. The significance of considering the approaches is even higher when we consider the nature of data sets with numerous associations and layers that we get from new and emerging media.

References

Bouckaert, R. R. (2004). Bayesian networks in Weka. *Technical Report 14/2004.* Computer Science Department, University of Waikato, 1–43.

Borsuk, M. E., Stow, C. A., & Reckhow, K. H. (2004). A Bayesian network of eutrophication models for synthesis, prediction, and uncertainty analysis. *Ecological Modelling, 173*(2), 219–239.

King, G. (2011). Ensuring the data-rich future of the social sciences. *Science, 331*(6018), 719–721.

Lazer, D., Pentland, A. S., Adamic, L., Aral, S., Barabasi, A. L., Brewer, D., … Jebara, T. (2009). Life in the network: The coming age of computational social science. *Science, 323*(5915), 721.

Marks, M. A., DeChurch, L. A., Mathieu, J. E., Panzer, F. J., & Alonso, A. (2005). Teamwork in multiteam systems. *Journal of Applied Psychology, 90*(5), 964.

Pe'er, D. (2005). Bayesian network analysis of signaling networks: A primer. *Sci STKE, 281,* 1–12.

Pilny, A., Yahja, A., Poole, M.S., & Dobosh, M. (2014). A dynamic social network experiment with multiteam systems. *Big Data and Cloud Computing, Proceedings of 2014 Social Computing,* 587–593.

Witten, I. H., Frank, E., & Hall, M. A. (2011). *Data mining: Practical machine learning tools and techniques* (3rd ed.,). San Francisco, CA: Morgan Kaufmann.

Chapter 4
A Relational Event Approach to Modeling Behavioral Dynamics

Carter T. Butts and Christopher Steven Marcum

4.1 Representing Interaction: From Social Networks to Relational Events

The social network paradigm is founded on the basic representation of social structure in terms of a set of social entities (e.g., people, organizations, or cultural domain elements) that are, at any given moment in time, connected by a set of relationships (e.g., friendship, collaboration, or association) (Wasserman & Faust, 1994). The success of this paradigm owes much to its flexibility: with substantively appropriate definitions of entities (*vertices* or *nodes* in network parlance) and relationships (*ties* or *edges*), networks can serve as faithful representations of phenomena ranging from communication and sexual relationships to neuronal connections and the structure of proteins (Butts, 2009). Nor must networks be static: the time evolution of social relationships has been of interest since the field's earliest days (see, e.g. Heider, 1946; Rapoport, 1949; Sampson, 1969), and considerable progress has been made on models for network dynamics (e.g. Snijders, 2001; Koskinen & Snijders, 2007; Almquist & Butts, 2014; Krivitsky & Handcock, 2014). Such models treat relationships (and, in some cases, the set of social entities itself) as evolving in discrete or continuous time, driven by mechanisms whose presence and strength can be estimated from intertemporal network data.

C.T. Butts (✉)
University of California Irvine, Irvine, CA, USA
e-mail: buttsc@uci.edu

C.S. Marcum
National Institutes of Health, Bethesda, MD, USA
e-mail: chris.marcum@nih.gov

© Springer International Publishing AG 2017   51
A. Pilny, M.S. Poole (eds.), *Group Processes*, Computational Social Sciences,
DOI 10.1007/978-3-319-48941-4_4

A key assumption that underlies the network representation in both its static and dynamic guises is that relationships are *temporally extensive*—that is, it is both meaningful and useful to regard individual ties as being present for some duration that is at least comparable to (and possibly much longer than) the time scale of the social process being studied. Where tie durations are much longer than the process of interest, we may treat the network as effectively "fixed;" thus is it meaningful for Granovetter (1973) or Burt (1992) to speak of personal ties as providing access to information or employment opportunities, for Freidkin (1998) to model opinion dynamics in experimental groups, or for Centola and Macy (2007) to examine the features that allow complex contagions to diffuse in a community, without explicitly treating network dynamics. When social processes (including tie formation and dissolution themselves) occur on a timescale comparable to tie durations, it becomes vital to account for network dynamics. For instance, the diffusion of HIV through sexual contact networks is heavily influenced by partnership dynamics (particularly the formation concur rent rather than serial relationships) (Morris, Goodreau, & Moody, 2007), and health behaviors such as smoking and drinking among adolescents are driven by an endogenous interaction between social selection and social influence (see, e.g. Lakon, Hipp, Wang, Butts, & Jose, 2015; Wang, Hipp, Butts, Jose, & Lakon, 2016). While there are many practical and theoretical differences between the behavior of networks in the dynamic regime versus the "static" limit, both regimes share the common feature of *simultaneity:* relationships overlap in time, allowing for apparent reciprocal interaction between them.

Such simultaneous co-presence of edges forms the basis of all network structure (as expressed in concepts ranging from reciprocity and transitivity to centrality and structural equivalence), and is the foundation of social network theory. Such simultaneity, however, is a hidden consequence of the assumption of temporal extensiveness; in the limit, as tie durations become much shorter than the timescale of relationship formation, we approach a regime in which "ties" become fleeting interactions with little or no effective temporal overlap. In this regime the usual notion of network structure breaks down, while alternative concepts of *sequence* and *timing* become paramount.

This regime of social interaction is the domain of *relational events* (Butts, 2008). Relational events, analogous to edges in a conventional social network setting, are discrete instances of interaction among a set of social entities. Unlike temporally extensive ties, relational events are approximated as instantaneous; they are hence well-ordered in time, and do not form the complex cross-sectional structures characteristic of social networks. This lack of cross-sectional structure belies their richness and flexibility as a representation for interaction dynamics, which is equal to that of networks within the longer-duration regime. (In fact, the two regimes can be brought together by treating relationships as spells with instantaneous start and end events. Our main focus here is on the instantaneous action case, however). The relational event paradigm is particularly useful for studying the social action that lies beneath (and evolves within) ongoing social relationships. In this settings, relational events are used to represent particular instances of social behavior (e.g., communication, resource transfer, or hostility) exchanged between individuals. To understand how such behaviors unfold over time requires a theoretical framework and analytic foundation that incorporates the distinctive properties of such micro-behaviors. Within the relational event paradigm, actions (whether individual or collective) are treated as arising as discrete events in continuous time, whose hazards are potentially complex functions of the properties of the actors, the social context, and the history of prior interaction itself (Butts, 2008). In this way, the relational event paradigm can be viewed as a fusion of ideas from social networks and allied theoretical traditions such as agent-based modeling with the inferential foundation of survival and event history analysis (Mayer & Tuma, 1990; Blossfeld & Rohwer, 1995). The result is a powerful framework for studying complex social mechanisms that can account for the heterogeneity and context dependence of real-world behavior without sacrificing inferential tractability.

4.1.1 Prefatory Notes

At its most elementary level, as Marcum and Butts (2015) point out, the relational event framework helps researchers answer the question of "what drives what happens next" in a complex sequence of interdependent events. In this chapter, we briefly review the relational event framework and basic model families, discuss issues related to data selection and preparation, and demonstrate relational event model analysis using the freely available software package *relevent* for R (Butts, 2010). Here, we provide some additional context before turning to the data and tutorial.

Following Butts (2008), a relational event is defined as an action emitted by one entity and directed toward another in its environment (where the entities in question may be sets of more primitive entities (e.g., groups of individuals), and self-interactions may be allowed). From this definition, a relational event is thus comprised of a sender of action, a receiver of that action, and a type of action, with the action occurring at a given point in time. In the context of a social system, we consider relational events as "atomic units" of social interaction. A series of such events, ordered in time, comprise an event history that records the sequence of social actions taken by a set of senders and directed to a set of receivers over some window of observation. The set of senders and the set of receivers may consist of human actors, animals, objects or a combination of different types of actors. The set of action types, likewise, may consist of a variety behaviors including communication, movements, or exchanges.

The relational event framework is in an increasingly popular approach to the analysis of relational dynamics and has been adopted by social network researchers in a wide variety of fields. Typically, research questions addressed in this body of work focus on understanding the behavioral dynamics of a particular type of action (such as communication alone).

Recently, relational event models have been used to study phenomena as diverse as reciprocity in food-sharing among birds (Tranmer, Marcum, Morton, Croft, & de Kort, 2015); social disruption in herds of cows (Patison, Quintane, Swain, Robins,

& Pattison, 2015); cooperation in organizational networks (Leenders, Contractor, & DeChurch, 2015); conversational norms in online political discussions (Liang, 2014); and multiple event histories from classroom conversations (DuBois, Butts, McFarland, & Smyth, 2013b).

Prior to the relational event framework, behavioral dynamics occurring within the context of a social network were generally modeled using frameworks developed for dynamic network data; since, as noted above, dynamic networks are founded on the notion of simultaneous, temporally extensive edges, use of dynamic network models for relational event data requires aggregation of events within a time window. Such aggregation leads to loss of information, and the results of subsequent analyses may depend critically on choices such as the width of the aggregation window. Model families such as the stochastic actor-oriented models (Snijders, 1996) or the temporal exponential random graph models (Robins & Pattison, 2001; Almquist & Butts, 2014; Krivitsky & Handcock, 2014) are appropriate for studying systems of simultaneous relationships that evolve with time, but may yield misleading results when fit to aggregates of relational events. While such use can be motivated in particular cases, we do not as a general matter recommend coercing event processes into dynamic network form for modeling purposes. Rather, where possible, we recommend that relational event processes be treated on their own terms, as sequences of instantaneous events with relational structure. In the following sections, we provide an introduction to this mode of analysis.

4.2 Overview of the Relational Event Framework

We begin our overview of the relational event framework by considering what a relational event process entails. Although we provide some basic notation, we omit most technical details; interested readers are directed to Butts (2008), DuBois et al. (2013b), and Marcum and Butts (2015) for foundations and further developments. We start with a set of potential senders, S, a set of potential receivers, R, and a set of action types, C. A "sender" or "receiver" in this context may refer to a single individual or a set thereof; in some cases, it may be useful to designate a single bulk sender or receiver to represent the broader environment (if, e.g., some actions may be untargeted, or may cross the boundary between the system of interest and the setting in which that system is embedded). An example of the use of aggregate senders and receivers is shown in Sect. 4.3.1. A single action or relational event, a, is then defined to be a tuple containing the sender of that action $s = s(a) \in S$, the receiver of the action $r = r(a) \in R$, the type of action $c = c(a) \in C$, and the time that the action occurred $\tau = \tau(a)$; formally, $a = (s, r, c, t)$, the analog of an edge in a dynamic network setting. In practice, we may associate one or more covariates with each potential action (X_a), relating to properties of the sender or receiver, the sender/receiver dyad, the time period in question, et cetera. A series of relational events observed from time 0 (defined to be the onset of observation) and a certain time t comprise an event history, denoted $A_t = \{A_i: \tau(a_i) <= t\}$. Typically, we will observe a realization of At and seek to infer the mechanisms that generated (which will be expressed via a set of parameters, θ, as described below). At any given point in the event history, the set of possible events (or *support*) is defined by the set $\mathbf{A}(A_t) \subseteq S \times R \times C$, where $\times$ indicates the Cartesian product. We note that the support may be endogenous, allowing us to consider cases in which particular actions within the event history either make new actions possible or render previously available actions impossible, or exogenous whereby certain possibilities in the support have been restricted (or otherwise new opportunities availed) due to circumstances outside of the system under study. (For instance, an individual who has left a building cannot speak to those still within it, and the appearance of a new entrant provides a new target for interaction).

Let $\mathbf{A}$ define the set of events that are possible at any moment. The propensity of such an event to occur is defined via its *hazard*, i.e. the limit of the conditional rate of event occurrence in a time window about a given point, as the width of that window approaches 0. Intuitively, the hazard of relational event a at time t is non-negative and equal to 0 if and only if $a \notin \mathbf{A}(A_t)$ (i.e., a is currently impossible); larger hazards correspond to higher propensities. It is important to note that each event that is possible at a given moment has a non-zero hazard, and not merely those events that happen to occur; by observing both the events that transpired and the events that could have transpired (but did not), we seek to infer the propensities for all possible events. Such inference requires that we parameterize our event hazards, and it is natural to conceive of each as arising from a combination of mechanistic factors that either enhance or inhibit the realization of the event in question. Typically, we implement this by asserting that the hazard of each event is a multiplicative function of a series of statistics, each of which encodes the effect of a given mechanism on event propensity. Formally, this is expressed (Eq. 4.1) as:

$$\lambda_{aAt\theta} = \begin{cases} \exp(\theta^T u(s(a),r(a),c(a),X_a A_t)) & \text{if } a \in \mathbb{A}(A_t), \\ 0 & \text{otherwise} \end{cases} \tag{4.1}$$

where $\lambda_{aAt\theta}$ is the hazard of potential event a at time t given history A_t, θ is a vector of real-valued parameters, and u is a vector of functions (i.e., statistics) that may depend upon the sender, receiver, or type of an event, covariates, and/or the prior event history. It should be noted that the log-linear form for the hazard function used above is not strictly necessary, and other forms are possible. However, we do not consider such alternatives here.

The role played by the u functions in a relational event model is analogous to that of the sufficient statistics in an exponential random graph model (see, e.g. Wasserman & Robins, 2005), or to the effects in a conventional hazard model (Blossfeld & Rohwer, 1995): each represents a mechanism that may increase or decrease the propensity of a given action to be taken, as governed by θ. Each unit change in u_i multiplies the hazard of an associated event by $\exp(\theta_i)$, thereby making it (ceteris paribus) more prevalent and quick to occur or less prevalent and slower to occur. Typically, candidates for u are proposed on a priori theoretical grounds, with θ being inferred from available data. Comparison of goodness-of-fit for models with

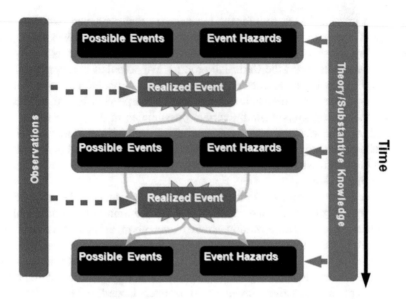

Fig. 4.1 Schematic representation of the inferential logic of the relational event framework. Models, proposed on theoretical grounds, determine the set of possible events and the mechanisms governing event hazards; observations of realized events are employed to infer unknown parameters governing the strengths and directions of effects, and to select among competing models

alternative choices of u allows for alternative theories of social mechanisms to be tested, without assuming that the dynamics are governed by any single mechanism.

Figure 4.1 illustrates the logic of relational event framework by depicting a very general relational event process together with its theoretical components. In this figure, time runs downward from the top of the illustration to the bottom (as indicated by the rightmost vertical axis). We begin with the state of the world *prior to* any observation of a relational event. This state can be characterized by the set of potential actions (or possible events) and their underlying propensities to occur (or their respective event hazards). For example, we may observe a group of individuals in a room, each of whom may direct a speech act at the others, with the hazards representing the distribution of action propensities. Then, something happens: we observe a realized relational event—one of the actors (the sender) addresses another actor (the receiver). The occurrence of this particular action, in turn, may have changed the state of the world, possibly including what actions are possible and each individual's propensity to act. For instance, speaking first may have emboldened the first sender and incremented her propensity to speak even more. Thus, we update the set of possible events and their hazards to reflect new information given the current state of the event history. Next, something else happens: we observe another relational event. Again, this event may change the set of possible events and their hazards, and we update our view of the world based on the past history. This process continues by turns until the last event (not shown). Just as we make observations on the sequence of events, we use theory and substantive knowledge about the world to make suppositions or impose limits on the set of possible events and to derive the u statistics that govern the event hazards.

As the above indicates, the types of effects we estimate using the relational event framework can capture a wide range of mechanisms involving both endogenous behavioral dynamics and exogenous effects (either covariate-based or the impact of exogenous events). Typical examples include actor-level fixed effects (rates for sending and receiving events for each actor), subsequence effects, and time invariant and time varying covariate effects. There are many possibilities for modeling endogenous dynamics using the relational event framework because there are many types of event history sub-sequences from which one may build sufficient statistics. Some such sub-sequences are of general theoretical interest. For example, we may consider the social processes related to the persistence of action, order of action, exchanges within triads of actors, conversational dynamics, or even dynamic preferential attachment. Each of these processes can be parameterized in terms of a series of prior events in the life history, allowing it to be implemented in the relational event framework. The selection of such effects to proposed in a candidate model should be driven by the research question and evaluated by assessing goodness-of-fit (options currently supported by software are listed in the tutorial, below). For example, much research has shown that persons who have interacted frequently in the past are likely to continue to interact in the future. In a relational event context, we might thus hypothesize that sending events to certain individuals increases the chances that they will remain the targets of events in the future. This behavior may be characterized as a type of social persistence or inertia and can be implemented with an effect that treats the fraction of previous contacts as a predictor of future contact. We might also hypothesize that the order in which one received ties from others in the past plays a role in one's likelihood of replying. Specifically, because the last thing that happened is very likely to be the most salient, we may model this process with a statistic that employs the inverse of the order of an actor's receipt of events from others as a predictor of that actor's sending of events back to them in the future. If the inclusion of this effect in the model substantially improves fit (net of degrees of freedom consumed), we conclude that the mechanism in question is predictive of the observed social process; if, however, we do not find such an improvement, we may thereby conclude that the observed pattern of interaction does not support the presence of the proposed mechanism. We return to more examples of relational event effects in the tutorial, below.

Regardless of which behaviors (or covariates) are of interest, it is important to understand the basic assumptions of the model used to estimate their effects on the relational event process; further details can be found in Butts (2008). Here, we briefly review three of the most relevant assumptions that most modelers should understand before using the relational event framework. First, we assume that all events are recorded, and that the onset of the observation period is exogenously determined (e.g., chosen by the researcher or set by a random external event). Second, we assume that no events can occur at exactly the same time but, rather, are temporally ordered. This assumption is perhaps the key distinction that separates the relational event regime from the dynamic network regime (as discussed above).

Finally, we typically assume that event hazards and the support are piecewise constant, with changes occurring either when an endogenous event is realized or at exogenous "clock" events. This final assumption has numerous useful implications, among them being ease of computation and interpretation, the ability to infer parameters when exact times are unknown, and the fact that the waiting times between events are conditionally exponentially distributed. (Piecewise constancy is also a standard assumption in the well-known Cox proportional hazards models (Mills, 2011), where it yields similar advantages). While this last assumption can be relaxed, current software implementations of the relational event framework (e.g. the *relevent* package for R, Butts, 2010) employ it.

Of these assumptions, the most critical is the notion that events are well-ordered in time. While non-simultaneity is in practice vital only for events whose occurrence can affect each others' hazards, and while there are various workarounds for data sets with small amounts of simultaneity (e.g., due to imprecise coding event times), large numbers of simultaneous events suggest a system which is not in the relational event regime. Such cases may be better represented as dynamic networks, in the manner discussed above.

While the relational event paradigm is defined in terms of instantaneous events that unfold in continuous time, inference for relational event models does not necessarily require that event times be known. It is useful in this regard to distinguish two general cases: event histories in which only the order of events is known ("ordinal time"); and event histories in which the exact time between events is known ("exact" or "interval time"). Butts (2008) derives the model likelihood for both scenarios under the assumptions listed above. Importantly, under the assumption of piecewise constant hazards, the parameter vector θ can in principle be identified up to a pacing constant; since relative rather than absolute hazards are typically of primary scientific interest, this implies that information on event ordering is frequently adequate to employ the framework. Such data is common e.g. in archival or observational settings, in which it may be feasible to construct a transcript of actions taken but difficult or impossible to time them accurately. Both the ordinal and exact cases can be analyzed using the *relevent* package which, supports a variety of model effects. Additionally, while we are here focused on the basic case dyadic relational event models in a single event history, the framework is general enough to accommodate multiple event histories and even ego-centered event histories (DuBois et al., 2013b; Marcum & Butts, 2015) should one possess those types of data.

4.3 Sample Cases

To illustrate the use of the relational event model (REM) family, we employ sample case data from two previously published sources. First, to illustrate the relational event model for ordinal time data, we use data from Butts, Petrescu-Prahova, and Cross (2007). These data consist of radio communications among 37 named communicants in a police unit that responded to the World Trade Center disaster on the morning of September 11th, 2001. Second, to illustrate

REMs for exactly timed data, we use a time-modified[1] subset of data from Dan McFarland, who recorded conversations occurring between 20 participants in classroom discussions (Bender-deMoll & McFarland, 2006). Both datasets are available online for didactic purposes here.

For the *relevent* software package used in the tutorial below, data are stored in "rectangular" format as an $m \times 3$ matrix we call an "edgelist" (where m is the number of events). The first column of the edgelist indexes either the time or the order of the events, depending on the type of data. The second and third columns index the senders and receivers of the events, respectively, numbered from 1 to n (where n is the number of interacting parties). Importantly, the edgelist must be ordered by the first column (i.e., by time or event order). For exact timing data, the last row of the edgelist should index a null event for the time that observation period ended (by default, any event occurring in this row will be ignored by the software).

Optional sender and receiver covariate data may be stored separate from the edgelist as vectors or arrays, provided that they are ordered consistently with the actor set (1 through the number of actors). For time invariant covariates, this will be an $n \times p$ matrix, where n indexes the actors and p indexes the covariates. For time varying actor covariates, data should be stored in a 3-dimensional $m \times p \times n$ array, where m indexes time and p and n index covariates and actors as above.

Optional event covariate data may be stored similarly. For time invariant covariates, the data should be stored in a 3-dimensional $p \times n \times n$ array, where p and n index each fixed covariate and actor, respectively. Likewise, time varying event covariates should be stored in a 4-dimensional $m \times p \times n \times n$ array, where m indexes time and the other dimensions are as above.

4.3.1   Butts et al.'s WTC Data

The 9/11 terrorist attacks at the World Trade Center (WTC) in New York City in 2001 set off a massive response effort, with police being among the most prominent responders. As in much routine police work, radio communication was essential in coordinating activities during the crisis. Butts et al. (2007) coded radio communication events between officers responding to 9/11 from transcripts of communications recorded during the event. We will illustrate ordinal time REMs using the 481 communication events from 37 named communicants in that data set. It is important to note that the WTC radio data was coded from transcripts that lacked detailed timing information; we do not therefore know precisely when these calls were made. We do, however, know the order in which calls were made, and can use this to fit temporally ordinal relational event models. Additionally, we will employ a single actor-level covariate from this dataset: an indicator for whether or not a communicant filled an institutional coordinator role, such as a dispatcher (Petrescu-Prahova & Butts, 2008).

---

[1] Some events were given in order, but not distinguished by time; these have been spaced by 0.1 min for purposes of illustration.

4.3.2   McFarland's Classroom Data

Dan McFarland's classroom dataset includes exactly timed interactions between students and instructors within a high school classroom (McFarland, 2001; Bender-deMoll & McFarland, 2006). Sender and receiver communication events (n=691) were recorded between 20 actors (18 students and 2 teachers) along with the time of the events in increments of minutes. The data employed here were modified slightly to increase the amount of time occurring between very closely recorded events, ensuring no simultaneity of events as assumed by the relational event framework. Two actor-level covariates are also at hand in the dataset used here: whether the actor was a teacher and whether the actor was female.

4.4   Tutorial

Software for fitting relational event models is provided by the *relevent* package for R (Butts, 2010). There are numerous tutorials available online that provide instruction on how to obtain and learn to use the free R software. We direct neophyte users to the R project website (CRAN) to browse those resources: https://cran.r-project.org/. In this tutorial we assume that R is installed and users have some experience with statistical programming in that environment.

The *relevent* package and its dependencies can be downloaded from CRAN using R, installed, and loaded into the user's environment in the usual manner:

```
> install.packages("relevent")
> library(relevent)

> load("remdata.Rdata")
> ls( )
 [1] "as.sociomatrix.eventlist" "Class"
 [3] "ClassIntercept" "ClassIsFemale"
 [5] "ClassIsTeacher" "sleep.glbs"
 [7] "sleep.int" "wtc.coord"
 [9] "WTCPoliceCalls" "WTCPoliceIsICR"
[11] "WTCPoliceNet"
```

Dyadic relational event models are intended to capture the behavior of systems in which individual social units (persons, organizations, animals, etc.) direct discrete actions towards other individuals in their environment. Within the *relevent* package, the rem.dyad() function is the primary workhorse for modeling dyadic data. From the supplied documentation in R, the rem.dyad() function definition lists a number of arguments and parameters:

rem.dyad(edgelist, n, effects = NULL, ordinal = TRUE, acl = NULL,

cumideg = NULL, cumodeg = NULL, rrl = NULL, covar = NULL, ps = NULL,

tri = NULL, optim.method = "BFGS", optim.control = list ( ),

coef.seed = NULL, hessian = FALSE, sample.size = Inf, verbose = TRUE,

fit.method = c ("BPM", "MLE", "BSIR"), conditioned.obs = 0,

prior.mean = 0, prior.scale = 100, prior.nu = 4, sir.draws = 500,

sir.expand = 10, $sir.nu$ = 4, gof = TRUE)

In this tutorial, we focus on the first four arguments—*edgelist, n, effects, ordinal*; the ninth argument *covar*; and the fifteenth argument hessian. The remaining arguments govern model fitting procedures and output and their default values will suffice here. The first argument, *edgelist*, is how the user passes their data to rem. dyad; aptly, this takes an edgelist as described above. The second argument, *n*, should be a single integer representing the number of actors in the network. The third argument, *effects*, is how the user specifies which statistics (effects) will be used to model the data. This argument should be a character vector where each element is one or more of the following pre-defined effect names:

- 'NIDSnd': Normalized indegree of v affects v's future sending rate
- 'NIDRec': Normalized indegree of v affects v's future receiving rate
- 'NODSnd': Normalized outdegree of v affects v's future sending rate
- 'NODRec': Normalized outdegree of v affects v's future receiving rate
- 'NTDegSnd': Normalized total degree of v affects v's future sending rate
- 'NTDegRec': Normalized total degree of v affects v's future receiving rate
- 'FrPSndSnd': Fraction of v's past actions directed to v' affects v's future rate of sending to v'

- 'FrRecSnd': Fraction of v's past receipt of actions from v' affects v's future rate of sending to v'

- 'RRecSnd': Recency of receipt of actions from v' affects v's future rate of sending to v'

- 'RSndSnd': Recency of sending to v' affects v's future rate of sending to v'
- 'CovSnd': Covariate effect for outgoing actions (requires a 'covar' entry of the same name)

- 'CovRec': Covariate effect for incoming actions (requires a 'covar' entry of the same name)

- 'CovInt': Covariate effect for both outgoing and incoming actions (requires a 'covar' entry of the same name)

- 'CovEvent': Covariate effect for each (v,v') action (requires a 'covar' entry of the same name)

- 'OTPSnd': Number of outbound two-paths from v to v' affects v's future rate of sending to v'

- 'ITPSnd': Number of incoming two-paths from v' to v affects v's future rate of sending to v'

- 'OSPSnd': Number of outbound shared partners for v and v' affects v's future rate of sending to v'

- 'ISPSnd': Number of inbound shared partners for v and v' affects v's future rate of sending to v'

- 'FESnd': Fixed effects for outgoing actions
- 'FERec': Fixed effects for incoming actions
- 'FEInt': Fixed effects for both outgoing and incoming actions
- 'PSAB-BA': P-Shift effect (turn receiving)—AB!BA (dyadic)
- 'PSAB-B0': P-Shift effect (turn receiving)—AB!B0 (non-dyadic)
- 'PSAB-BY': P-Shift effect (turn receiving)—AB!BY (dyadic)
- 'PSA0-X0': P-Shift effect (turn claiming)—A0!X0 (non-dyadic)
- 'PSA0-XA': P-Shift effect (turn claiming)—A0!XA (non-dyadic)
- 'PSA0-XY': P-Shift effect (turn claiming)—A0!XY (non-dyadic)
- 'PSAB-X0': P-Shift effect (turn usurping)—AB!X0 (non-dyadic)
- 'PSAB-XA': P-Shift effect (turn usurping)—AB!XA (dyadic)
- 'PSAB-XB': P-Shift effect (turn usurping)—AB!XB (dyadic)
- 'PSAB-XY': P-Shift effect (turn usurping)—AB!XY (dyadic)
- 'PSA0-AY': P-Shift effect (turn continuing)—A0!AY (non-dyadic)
- 'PSAB-A0': P-Shift effect (turn continuing)—AB!A0 (non-dyadic)
- 'PSAB-AY': P-Shift effect (turn continuing)—AB!AY (dyadic)

The fourth argument, *ordinal*, is a logical indicator that determines whether to use the ordinal or exact timing likelihood. The default setting specifies ordinal timing (TRUE). The ninth argument, *covar*, is how the user passes covariate data to rem.dyad(). Objects passed to this argument should take the form of an R list, where each element of the list is a covariate as described above. When covariates are indicated, then there should be an associated covariate effect listed in the effects argument and each element of the *covar* list should be given the same name as its corresponding effect type specified in *effects* (e.g., 'CovSnd', 'CovRec', etc).

Finally, the fifteenth argument *hessian* is a logical indicator specifying whether or not to compute the Hessian of the log-likelihood or posterior surface, which is used in calculating inferential statistics. The default value of this argument is FALSE.

Having introduced the relational event package and the model fitting function, we now transition to examples of fitting relational event models using the two data-sets described above. Since the case of ordinal timing is somewhat simpler than that of exact timing, we consider the World Trade Center data first in the tutorial.

4.4.1   Ordinal Time Event Histories

Before we move to the analysis of the WTC relational event dataset, it is useful to visually inspect both the raw data and the time-aggregated network. The *eventlist* is stored in an object called WTCPoliceCalls. Examining the first six rows of this data reveals that the data is a matrix with the timing information, source (i.e., the sender, numbered from 1 to 37), and recipient (i.e., the receiver, again numbered from 1 to 37) for each event (i.e., radio call):

> head (WTCPoliceCalls)
number source recipient
1 1 16 32
2 2 32 16
3 3 16 32
4 4 16 32
5 5 1 32
6 6 1 32

Thus, we can already begin to see the unfolding of a relational event process just by inspecting these data visually. First, we see that responding officer 16 called officer 32 in the first event, officer 32 then called 16 back in the second (which might be characterized as a local reciprocity effect or AB → BA participation shift (Gibson, 2003)). This was followed by 32 being the target of the next four calls, perhaps due to either some unobserved coordinator role that 32 fills in the commu-nication structure or due to the presence of a recency mechanism. Further visual inspection is certainly warranted here. We can use the included *sna* function as. sociomatrix.sna() to convert the eventlist into a valued sociomatrix, which we can then plot using gplot():

> WTCPoliceNet < *as.*sociomatrix.eventlist (WTCPoliceCalls,37)

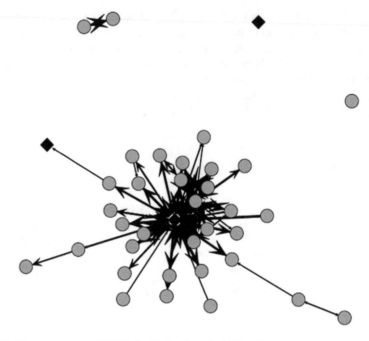

Fig. 4.2 Time-Aggregated WTC Police Radio Communication Network

> gplot(WTCPoliceNet, edge.*lwd* = WTCPoliceNet ^ .75, arrowhead.*cex* =
log(*as*.edgelist.*sna*(WTCPoliceNet)[,3]) + .25, vertex.*col* =
ifelse(WTCPoliceIsICR,,"black",,"gray"), vertex.*cex* = 1.25,
vertex.sides = ifelse(WTCPoliceIsICR,,4,,100), coord = *wtc*.coord)

Figure 4.2 is the resulting plot of the time-aggregated WTC Police communication network.

Your own may look slightly different due to both random node placement that gplot() uses to initiate the plot and because this figure has been tuned for printing. The three black square nodes represent actors who fill institutional coordinator roles and gray circle nodes represent all other communicants. A directed edge is drawn between two actors, i and j, if actor i ever called actor j on the radio. The edges and arrowheads are scaled in proportion to the number of calls over time. There are 37 actors in this network and the 481 communication events have been aggregated to 85 frequency weighted edges. This is clearly a hub-dominated network with two actors sitting on the majority of paths between all other actors. While the actor with the plurality of communication ties is an institutional coordinator (the square node at the center of the figure), heterogeneity in sending and receiving communication ties is evident, with several high-degree non-coordinators and two low-degree institutional coordinators, in the network. This source of heterogeneity is a good starting place from which to build our model.

4.4.2   A First Model: Exploring ICR Effects

We begin by fitting a very simple covariate model, in which the propensity of individuals to send and receive calls depends on whether they occupy institutionalized coordinator roles (ICR). We fit the model by passing the appropriate arguments to rem.dyad and summarize the model fit using the summary() function on the fitted relational event model object.

> wtcfit1 < rem.dyad(WTCPoliceCalls, n = 37, effects = c("CovInt"),
        covar = list(CovInt = WTCPoliceIsICR),
hessian = TRUE)
Computing preliminary statistics
Fitting model
Obtaining goodness of fit statistics
> summary(wtcfit1)
Relational Event Model(Ordinal Likelihood)
              Estimate $Std.Err$ Z value   Pr(> ||z||)
CovInt.1   2.104464 0.0698173 30.142 < 2.2e 16***

Signif.codes : 0 '***' '0.001 '**' 0.01 '*' '0.05.' 0.1 '1
Null deviance : 6921.048 on 481 degrees of freedom
Residual deviance : 6193.998 on 480 degrees of freedom
        Chi square : 727.0499 on 1 degrees of freedom,
              asymptotic p   value 0
AIC : 6195.998 AICC : 6196.007 BIC : 6200.174

The output gives us the covariate effect, as well as some uncertainty and goodness-of- fit information. The format is much like the output for a regression model summary, but coefficients should be interpreted per the relational event framework. In particular, the ICR role coefficient is the logged multiplier for the hazard of an event involving an ICR versus a non-ICR event ($e^{\lambda 1}$). This effect is cumulative: an event in which one actor in an ICR calls another actor in an ICR gets twice the log increment ($e^{2\lambda 1}$). We can see this impact in real terms as follows, respectively:

```
> # Relative hazard for a non ICR / ICR vs. a non ICR / non ICR
event
> exp(wtcfit1$coef)
CovInt.1
8.202707
> # Relative hazard for an ICR / ICR vs. a non ICR / non ICR event
    (twice the effect)
> exp(2 * wtcfit1$coef)
CovInt.1
67.28441
```

In this model, ICR status was treated as a homogenous effect on sending and receiving. Next, we evaluate whether it is worth treating these effects separately with respect to ICR status. To do so, we enter the ICR covariate as a sender and receiver covariate, respectively, and then evaluate which model is preferred by BIC (lower is better):

```
> wtcfit2 < rem.dyad(WTCPoliceCalls, n = 37, effects
    = c("CovSnd", "CovRec"),

covar = list(CovSnd = WTCPoliceIsICR, CovRec = WTCPoliceIsICR),
    hessian = TRUE)
Computing preliminary statistics
Fitting model


> summary(wtcfit2)
summary(wtcfit2)
Relational Event Model (Ordinal Likelihood)
            Estimate  Std.Err  Z value  Pr(> ||z||)
CovSnd.1  1.979175 0.095745 20.671  < 2.2e 16***
CovRec.1  2.225720 0.092862 23.968  < 2.2e 16***


Signif. codes : 0 '***' 0.001 '**' 0.01 '*' 0.05 '.' 0.1 ' ' 1
Null deviance : 6921.048 on 481 degrees of freedom
Residual deviance : 6190.175 on 479 degrees of freedom
        Chi square : 730.8731 on 2 degrees of freedom,
                asymptotic p value 0
AIC : 6194.175 AICC : 6194.2 BIC : 6202.527


> wtcfit1$BIC  wtcfit2$BIC
[1]  2.352663
```

While there appear to be significant ICR sender and receiver effects, their differences do not appear to be large enough to warrant the more complex model (as indicated by the slightly smaller Bayesian Information Criterion (BIC) of the first model). Smaller deviance-based information criteria, such as the BIC, indicate better model fit.

4.4.3   Bringing in Endogenous Social Dynamics

One of the attractions of the relational event framework is its ability to capture endogenous social dynamics. Next, we examine several mechanisms that could conceivably impact communication among participants in the WTC police network. In each case, we first fit a candidate model, then compare that model to our best fitting model thus far identified.

Where effects result in an improvement (as judged by the BIC), we include them in subsequent models, just as we decided for the comparison of the ICR covariate models.

To begin, we note that this is radio communication data. Radio communication is governed by strong conversational norms (in particular, radio standard operating procedures), which among other things mandate systematic turn-taking reciprocity. We can test for this via the use of what Gibson (2003) calls "participation shifts". In particular, the AB-BA shift, which captures the tendency for B to call A, given that A has just called B, is likely at play in radio communication. Statistics for these effects are described above. Building from our first preferred model, we now add this dynamic reciprocity term by including "PSAB-BA" in the effects argument to rem.dyad():

```
> wtcfit3 < rem.dyad(WTCPoliceCalls, n = 37,
effects = c("CovInt", "PSAB BA`),
    covar = list(CovInt = WTCPoliceIsICR), hessian = TRUE)
Computing preliminary statistics


Fitting model
Obtaining goodness of fit statistics
> summary(wtcfit3)
Relational Event Model( Ordinal Likelihood )
             Estimate Std.Err Z value    Pr(> ||z||)
CovInt.1   1.60405 0.11500 13.949 < 2.2e 16***
PSAB BA 7.32695 0.10552 69.436 < 2.2e 16***


Signif.codes : 0 *** '0.001 ** '0.01 * '0.05. 0.1 "1
Null deviance : 6921.048 on 481 degrees of freedom
Residual deviance : 2619.115 on 479 degrees of freedom
         Chi square : 4301.933 on 2 degrees of freedom,
                  asymptotic p value 0
AIC : 2623.115 AICC : 2623.14 BIC : 2631.467
> wtcfit1$BIC wtcfit3$BIC
[1] 3568.707
```

It appears that there is a very strong reciprocity effect and that the new model is preferred over the simple covariate model. In fact, the "PSAB-BA" coefficient indicates reciprocation events have more than 1500 times the hazard of other types of events ($e^{7.32695} = 1520.736$) that might terminate the AB—BX sub-sequence.

Of course, other conversational norms may also be at play in radio communication. For instance, we may expect that the current participants in a communication are likely to initiate the next call and that one's most recent communications may be the most likely to be returned. These processes can be captured with the participation shifts for dyadic turn receiving/continuing and recency effects, respectively:

> # Model 4 includes p shift effects
> wtcfit4 < rem.dyad(WTCPoliceCalls, n = 37, effects =
$$c\left("CovInt","PSAB\ BA","PSAB\ BY","PSAB\ AY"\right),$$
covar = list(CovInt = WTCPoliceIsICR), hessian = TRUE)
Computing preliminary statistics
Fitting model
Obtaining goodness of *fit* statistics

> summary(wtcfit4)
Relational Event Model(Ordinal Likelihood)
        Estimate *Std.Err* Z value   Pr(> ‖z‖)
CovInt.1 1.54283 0.11818 13.0549 < 2.2e 16***
PSAB *BA* 7.49955 0.11418 65.6831 < 2.2e 16***
PSAB *BY* 1.25941 0.25131 5.0115   5.402e 07***
PSAB *AY* 0.87215 0.30612 2.8491   0.004384**

Signif.codes : 0 '***' 0.001 '**' 0.01 '*' 0.05 '.' 0.1 ' ' 1
Null deviance : 6921.048 *on* 481 degrees of freedom
Residual deviance : 2595.135 *on* 477 degrees of freedom
     Chi square : 4325.913 *on* 4 degrees of freedom,
        asymptotic p value 0
AIC : 2603.135 AICC : 2603.219 *BIC* : 2619.839

> wtcfit3$*BIC* wtcfit4$*BIC*
[1] 12.62806

> # Model 5 adds recency effects *to* model 4
> wtcfit5 < *rem*.dyad(WTCPoliceCalls, n = 37,
                        effects = c("CovInt","PSAB *BA*","PSAB *BY*",
"PSAB *AY*","RRecSnd","RSndSnd"), covar =
                        list(CovInt = WTCPoliceIsICR), hessian = TRUE)
Computing preliminary statistics
Fitting model
Obtaining goodness of *fit* statistics

> summary(wtcfit5)
Relational Event Model(Ordinal Likelihood)
            Estimate *Std.Err* Z value Pr($> \|z\|$)
RRecSnd   2.38495 0.27447 8.6892 $< 2.2e$ 16***
RSndSnd   1.34623 0.22307 6.0350 1.590e 09***
CovInt.1   1.07058 0.14244 7.5160 5.640e 14***
PSAB *BA* 4.88714 0.15293 31.9569 $< 2.2e$ 16***
PSAB *BY* 1.67939 0.26116 6.4304 1.273e 10***
PSAB *AY* 1.39017 0.31057 4.4762 7.597e 06***

Signif.codes : 0 '***' 0.001 '**' 0.01 '*' 0.05 '.' 0.1 ' ' 1
Null deviance : 6921.048 *on* 481 degrees of freedom
Residual deviance : 2308.413 *on* 475 degrees of freedom
          Chi square : 4612.635 *on* 6 degrees of freedom,
                asymptotic p value 0
AIC : 2320.413 AICC : 2320.591 *BIC* : 2345.469
> wtcfit4$*BIC* wtcfit5$*BIC*
[1] 274.3701

The results indicate that turn-receiving, turn-continuing, and recency effects are all at play in the relational event process. Both models improve over the previous iterations by BIC, and the effect size reciprocity as been greatly reduced by controlling for other effects that reciprocity may have been masking in model 5 (i.e., the "PSAB-BA" coefficient was reduced from > 7 to > 4). Finally, recall that our inspection the time-aggregated network in Fig. 4.2 revealed a strongly hub-dominated

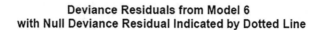

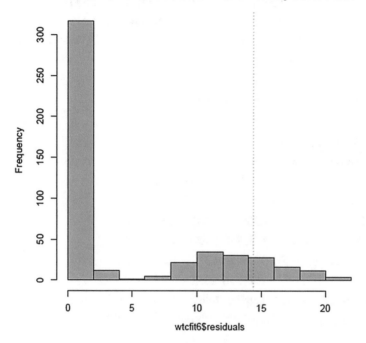

Fig. 4.3 Histogram of Deviance Residuals from Ordinal Model of WTC Data

network, with a few actors doing most of the communication. Could this be explained in part via a preferential attachment mechanism (per Price (1976) and others), in which those having the most air time become the most attractive targets for others to call? We can investigate this by including normalized total degree as a predictor of tendency to receive calls:

```
> wtcfit6 < rem.dyad(WTCPoliceCalls, n = 37,
            effects = c("CovInt", "PSAB BA", "PSAB BY",
"PSAB AY", "RRecSnd", "RSndSnd", "NTDegRec"),
            covar = list(CovInt = WTCPoliceIsICR), hessian = TRUE)
```

Computing preliminary statistics

Fitting model

Obtaining goodness of *fit* statistics

```
> summary( wtcfit6 )
```

Relational Event Model(Ordinal Likelihood)

	Estimate	Std.Err	Z value	Pr(> ‖z‖)	
NTDegRec	3.13453	0.56678	5.5305	3.194e 08	***
RRecSnd	2.02903	0.28500	7.1194	1.084e 12	***
RSndSnd	0.87116	0.23846	3.6533	0.0002589	***
CovInt.1	0.70734	0.16400	4.3129	1.611e 05	***
PSAB BA	5.32576	0.18236	29.2042	< 2.2e 16	***
PSAB BY	1.86023	0.26322	7.0673	1.579e 12	***
PSAB AY	1.64806	0.31092	5.3005	1.155e 07	***

Signif.codes : 0 '***' 0.001 '**' 0.01 '*' 0.05 '.' 0.1 ' ' 1

Null deviance : 6921.048 *on* 481 degrees of freedom

Residual deviance : 2277.263 *on* 474 degrees of freedom

Chi square : 4643.785 *on* 7 degrees of freedom,

asymptotic p value 0

AIC : 2291.263 AICC : 2291.5 *BIC* : 2320.494

```
> wtcfit5$BIC  wtcfit6$BIC
[1] 24.97434
```

Though still significant in the presence of preferential attachment effects, recency and ICR effect coefficients are reduced while participation shift effects are relatively unchanged. This final model is also preferred by BIC and it's clear that the deviance reduction from the null model is quite substantial at 67 %. While we could continue to investigate additional effects (see the list of options above), model 6 is a good candidate to evaluate model adequacy, which is addressed in the next section.

4.4.4 Assessing Model Adequacy

Model adequacy is an important consideration: even given that our final model from the exercises above (model 6) is the best of the set, is it good enough for our purposes? There are many ways to assess model adequacy; here, we focus on the ability of the relational event model to predict the next event in the sequence, given those that have come before. This approach nicely falls within the relational event framework. A natural question to ask in this framework is how "surprised" is the model by the data. Put another way, when does the model encounter relational event observations that are relatively poorly predicted? To investigate this, we can examine the deviance residuals, which are included in the fitted model object. We begin by calculating the deviance residual under the null which, from the ordinal likelihood derivation in Butts (2008), is simply twice the log product of the number of sender-receiver pairs, and comparing that with the deviance residuals under the fitted model:

```
> # Null deviance residual
> nullresid <   2*log(37*36)
> # Plot a histogram of the fitted model deviance residuals
> hist(wtcfit6$residuals, main = "Deviance Residuals from Model 6
\n with Null Deviance Residual Indicated", col = "gray")
> abline(v = nullresid, lty = 2)
> # What fraction are below the null resid?
> mean(wtcfit6$residuals < nullresid)
[1] 0.8898129
> # What fraction are less than 3?
> mean(wtcfit6$residuals < 3)
[1] 0.6839917
> # How "surprised" is the model?
mean(wtcfit6$residuals > nullresid)
[1] 0.1101871
```

The histogram of the model deviance residuals produced from the above code snippet is shown in Fig. 4.3. The dotted line indicates the null deviance residual: the idea here is that we want the model deviance residuals to fall to the right of that cut-off. Indeed, about 89 % of the model deviance residuals are smaller than the null residual, with 68 % of them being less than three (or really, really small). These initial checks are good conditional evidence that our model is performing really well.

To investigate further, we can evaluate the extent to which our model could take a random guess about which event comes next and get it right, relative to all possibilities. Here again, the deviance residuals come in handy as the quantity $e^{\frac{D}{2}}$, where D_i is the model deviance residual for event i, is a "random guessing equivalent". That is, it is the effective number of events such that a random guess about what happens next would be right as often as expected under the model.

```
> # Distribution of random guessing equivalents for model 6
> quantile (exp (wtcfit6$residuals / 2))
      0%           25%          50%          75%          100%
 1.073634     1.268661     1.739723    204.539040   31632.962350
> # Distribution of random guessing equivalents for model 1
> quantile (exp (wtcfit1$residuals / 2))
      0%           25%          50%          75%          100%
390.0003     390.0003     390.0003     390.0003     3199.0591
```

At least 50 % of the time our final model needs about 1 in 1.7 guesses to correctly predict the next event. This is in contrast to our first model with just the intercept term for ICR covariate, which needs about 390 such guesses. For an overall comparison, consider that the null model would get only 1 out of every 1332 (36 * 37) events correct just guessing at random.

Model adequacy as measured by surprise can also be visually inspected. First, one can inspect which events are surprising by adding an indicator for model surprise to the original eventlist:

```
> head (cbind (WTCPoliceCalls,surprise = wtcfit6$residuals > nullresid))
  number source recipient surprise
1    1      16      32     FALSE
2    2      32      16     FALSE
3    3      16      32     FALSE
4    4      16      32     FALSE
5    5      11      32     TRUE
```

The code snippet prints just the first five events, but these are enough to get a glimpse into why the model might be surprised. We can see that the first four events, involving exchanges between actors 16 and 32, are not surprising and appear to involve reciprocity and turn continuing participation shifts. The fifth event, however, is surprising, probably because it involves the sudden interruption of a new caller (actor 11). Thus, it appears that the model is surprised, perhaps unsurprisingly, when events transpire that are not specified by the model statistics such

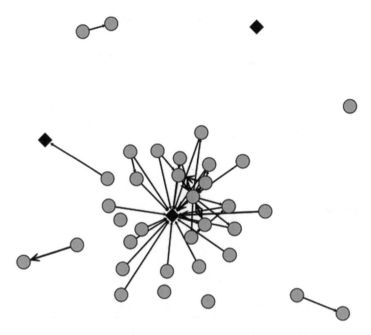

Fig. 4.4 Time-Aggregated 'Surprising' Events Network Under the Final Relational Event. Model of WTC Radio Communications

as third-party effects. These surprising events can also be projected onto the time-aggregated network using as.sociomatrix.sna, as before:

> surprising < *as.*sociomatrix.eventlist
$$\left(\text{WTCPoliceCalls}\left[\text{wtcfit6\$residuals} > \text{nullresid,}\right],37\right)$$

> gplot(surprising, edge.*lwd* = surprising ^.75,
$$\text{arrowhead.}cex = \log\left(as.\text{edgelist.}sna\left(\text{surprising}\right)\left[,3\right]\right)$$
$$+.25, \text{vertex.}col = \text{ifelse}\left(\text{WTCPoliceIsICR,,"black",,"gray"}\right),$$
$$\text{vertex.}cex = 1.25, \text{vertex.sides} =$$
$$\text{ifelse}\left(\text{WTCPoliceIsICR,,,,100}\right)\right)$$

The resulting plot of the time-aggregated surprising event network is illustrated in Fig. 4.4, which can be directly compared with Fig. 4.2. While there are many fewer events that are surprising than not, it's clear from the figure that the surprising events resolve on where the greatest opportunity for communication exists: namely on calls directed toward the main hub at the center and also calls sent from the sec-ondary hub to others. This suggests the existence of some unobserved heterogeneity related to those actors not explained by conversational norms, preferential attach-ment to them, or whether or not they fill institutional coordinator roles.

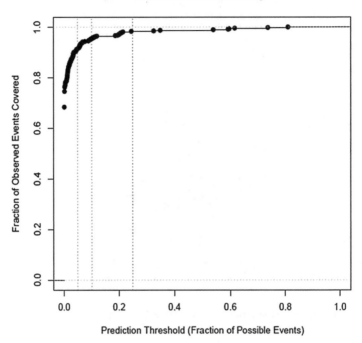

Classification Accuracy

Fig. 4.5 Classification Accuracy of the Observed Ranks Under Model 6 with Prediction. Thresholds Indicated at 0.05, 0.1, and 0.25

Finally, the function rem.dyad() supplies two additional components in returned model objects that are useful for evaluating adequacy. These are the rank of the observed events in the predicted rate structure and a pair of indicators for whether or not the model exactly predicts the sender and receiver, respectively, involved in each event. While far more stringent as measures of surprise than the deviance residuals, these statistics can be quite informative for well-fitting models.

For instance, we can inspect the empirical cumulative distribution function of the observed ranks to assess classification accuracy of the model at various thresholds:

> plot(ecdf $\left(\text{wtcfit6\$observed.rank} / \left(37 * 36\right)\right)$, xlab = "Prediction Threshold

$\left(\text{Fraction of Possible Events}\right)$", ylab = "Fraction of Observed Events Covered", main

= "Classification Accuracy")

> abline $\left(v = c\left(0.05,,0.1,,0.25\right), lty = 2\right)$

The resulting plot of the ECDF is shown in Fig. 4.5, which shows that predictions under the model very quickly cover the observed events. For the strictest measures, we can ask three questions of the exact predictions: (1) what is the fraction of events for which either sender or receiver are exactly predicted; (2) what is the fraction of events for which both sender and receiver are exactly predicted; and, (3) what are the respective fractions of events where we get the sender and receiver right under the model. These questions are easily addressed using the fitted model output:

> mean$\left(\text{apply}\left(\text{wtcfit6\$predicted.match},1,any\right)\right)$

[1]0.7941788

> mean$\left(\text{apply}\left(\text{wtcfit6\$predicted.match},1,all\right)\right)$

[1]0.6839917

> colMeans$\left(\text{wtcfit6\$predicted.match}\right)$

    source   recipient

0.7234927 0.7546778

Thus, our final model predicts something right about 79 % of the time (getting the sender right for 72 % and the receiver right about 75 % of the events, respectively) and it predicts the event that actually transpired exactly right 68 % of the time. Despite its simplicity, this model appears to fit extremely well. Further improvement is possible, but for many purposes we might view it as an adequate representation of the event dynamics in this WTC police radio communication network.

4.5   Exact Time Histories

We now turn to a consideration of REMs for event histories with exact timing information. As in the case of ordinal time data, it is useful to begin by examining the raw temporal data and the time-aggregated network. The edgelist is stored in an object called Class. Printing the first six rows and the last two rows of this object reveals minor differences between the exact time and the ordinal time data structures (discussed above). As before, we have three columns: the event time, the event source (numbered from 1 to 20), and the event target (again, numbered 1 to 20). In this case, event time is given in increments of minutes from onset of observation. Note that the last row of the event list contains the time at which observation was terminated; it (and only it) is allowed to contain NAs, since it has no meaning except to set the period during which events could have occurred. Where exact timing is used, the final entry in the edgelist is always interpreted in this way, and any source/target information on this row is ignored. This row indicates that the total period of observation lasted just over 50 minutes (the length of one class session).

Fig. 4.6 Time-Aggregated
Classroom Communications

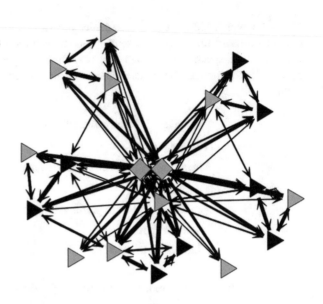

$$> \mathrm{Class}\big[\mathrm{c}\,(1:6,691:692),\big]$$

	StartTime	FromId	ToId
1	0.135	14	12
2	0.270	12	14
3	0.405	18	12
4	0.540	12	18
5	0.675	1	12
6	0.810	12	1
691	50.910	17	6
692	50.920	*NA*	*NA*

We can again use the sna toolkit to convert and plot the time-aggregated network
for inspection. Here, we color the female nodes black and the male nodes gray and
represent teachers as square-shaped nodes and students as triangle-shaped nodes.
Edges between nodes are likewise scaled proportional to the number of communication events transpiring between actors.

```
ClassNet < as.sociomatrix.eventlist (Class,20)
gplot(ClassNet, vertex.col = ifelse (ClassIsFemale,, "black ",, "gray "), vertex.sides = 3 +
ClassIsTeacher, vertex.cex = 2, edge.lwd = ClassNet ^ .75)
```

Figure 4.6 displays the resulting time-aggregated network. A dynamic visualization of this data is also available online in (Bender-deMoll & McFarland, 2006) and
is well worth examining. While it is clear from this figure that teachers do a great

deal of talking, there also appear to be several high-degree students. Female students in this classroom also appear to be slightly more peripheral. Both of these observations warrant inclusion of the respective covariates in our analysis, to which we now turn.

4.5.1   Modeling with Covariates

One of the advantages that the exact time relational event model likelihood has over the ordinal time likelihood is its ability to estimate pacing constants (i.e., the global rates at which events transpire). Here we investigate this with a simple intercept model, containing only a vector of 1 s as an actor-level sending effect. This vector is saved as ClassIntercept, which we can pass to the respective covariate arguments in rem.dyad(). Note that we must also tell rem.dyad that we do not want to discard timing information by setting the argument ordinal=FALSE:

> $\text{classfit1} < \ rem.\text{dyad}(\text{Class}, \text{n} = 20, \text{effects} = c(\text{"CovSnd"}),$
$\quad\quad\quad \text{covar} = \text{list}(\text{CovSnd} = \text{ClassIntercept}),$
$\text{ordinal} = \text{FALSE}, \text{hessian} = \text{TRUE})$
Computing preliminary statistics
Fitting model
Obtaining goodness of *fit* statistics

> $\text{summary}(\text{classfit1})$
Relational Event Model (Temporal Likelihood)
$\quad\quad\quad\quad\quad$ Estimate   *Std.Err*   Z value $\text{Pr}(> \|z\|)$
CovSnd.1   3.332287   0.038042   87.596 $< 2.2e \ 16^{***}$

Signif.codes : 0^{***}'0.001^{**}'0.01^{*}'0.05'.'0.1'"1
Null deviance : $5987.221 \, on \, 691$ degrees of freedom
Residual deviance : $5987.221 \, on \, 691$ degrees of freedom
$\quad\quad$ *Chi* square : $4.274625e \ 11 \, on \, 0$ degrees of freedom,
$\quad\quad\quad$ asymptotic p value 1
$\quad$ *AIC* : 5989.221 AICC : 5989.227 *BIC* : 5993.759

The model does not fit any better than the null because it is equivalent to the null model (as indicated by the absence of difference between the null and residual deviance). As one would expect from first principles, this is really just an exponential waiting time model, calibrated to the observed communication rate. Thus, to calculate the predicted number of events per minute we may multiply the number of possible event types (here, $20 \_ 19 = 380$) by the coefficient for the intercept:

> 380 * exp (classfit1 $coef)

CovSnd.1

13.57031

This simple model predicts the overall pace of events to occur at nearly 14 events per minute and this matches quite well with the average number of events per minute from the observed data:

$$> (\text{nrow}(\text{Class})\ 1)/\max(\text{Class}[,1])$$

$$[1]13.57031$$

Because we noted structural heterogeneity based on gender and status in Fig. 4.6, we fit a more interesting covariate model that specifies these effects for senders and receivers and evaluate whether there is any improvement over the intercept-only model by BIC.

> classfit2 < rem.dyad(Class, n = 20, effects = c ("CovSnd","CovRec"),

covar = list(CovSnd = cbind (ClassIntercept,,ClassIsTeacher,,ClassIsFemale),

CovRec = cbind (ClassIsTeacher,ClassIsFemale)), ordinal = FALSE, hessian = TRUE)

Computing preliminary statistics

Fitting model

Obtaining goodness of fit statistics

>

> summary (classfit2)

Relational Event Model (Temporal Likelihood)

	Estimate	Std.Err	Z value	Pr (> ‖z‖)	
CovSnd.1	3.834229	0.078842	48.6319	< 2e 16	***
CovSnd.2	1.672561	0.091679	18.2436	< 2e 16	***
CovSnd.3	0.123900	0.094931	1.3052	0.19184	
CovRec.1	0.373733	0.127028	2.9421	0.00326	**
CovRec.2	0.165729	0.080896	2.0487	0.04049	*

Signif.codes : 0 '***' 0.001 '**' 0.01 '*' 0.05 '.' 0.1 ' '1

Null deviance : 5987.221 on 691 degrees of freedom

Residual deviance : 5652.318 on 687 degrees of freedom

Chi square : 334.9034 on 4 degrees of freedom,

asymptotic p value 0

AIC : 5662.318 $AICC$: 5662.405 BIC : 5685.008

>

> classfit1 BIC classfit2 BIC

$[1]308.7508$

With multiple covariates, the model terms (*CovSnd.*1, *CovSnd.2* etc) are listed in the object in the same order as they were specified within the covar argument. Here, we see a good improvement over the null model but also note that gender does not appear to be predictive of sending communication. A better model may be one without that specific term included, which we fit below and again compare to the previous model by BIC.

> classfit3 < *rem*.dyad(Class, n = 20, effects = c ("CovSnd","CovRec"),

covar = list(CovSnd = cbind (ClassIntercept,ClassIsTeacher),

CovRec = cbind (ClassIsTeacher,ClassIsFemale)), ordinal = FALSE, hessian = TRUE)

Computing preliminary statistics

Fitting model

Obtaining goodness of *fit* statistics

> summary (classfit3)

Relational Event Model (Temporal Likelihood)

	Estimate	Std.Err	Z value	Pr(> ‖z‖)
CovSnd.1	3.775227	0.063623	59.3379	< 2.2e 16 ***
CovSnd.2	1.615762	0.079933	20.2139	< 2.2e 16 ***
CovRec.1	0.371749	0.127020	2.9267	0.003426 **
CovRec.2	0.161154	0.080815	1.9941	0.046141 *

Signif.codes : 0 *** '0.001 ** '0.01 * '0.05. 0.1 '1

Null deviance : 5987.221 *on* 691 degrees of freedom

Residual deviance : 5654.016 *on* 688 degrees of freedom

Chi   square : 333.2049 *on* 3 degrees of freedom,

asymptotic p   value 0

AIC : 5662.016 AICC : 5662.074 *BIC* : 5680.169

> classfit2$*BIC*   classfit3$*BIC*

[1] 4.839661

Indeed, there is a marginal improvement in BIC and we retain the model lacking the gender effect for sending communication events.

4.5.2 Modeling Endogenous Social Dynamics

While we find that the above covariate models perform better than the null, the final model is still unimpressive in terms of deviance reduction, with only about a 5 % total reduction from the null by our best fitting model. To investigate further, we propose a set of models that capture endogenous social dynamic effects that are reasonably presumed to be at play in classroom conversations. These include recency effects and effects that capture aspects of conversational norms, such as turn-taking, sequential address, and turn-usurping.

As before, we can enter these terms into the model using their appropriate effect names.

We also preserve the covariates from best covariate model (model 3 from the previous section) and check our improvement by BIC.

```
> # First, just recency effects + model 3 :
> classfit4 < rem.dyad(Class, n = 20,
          effects = c("CovSnd","CovRec","RRecSnd","RSndSnd"),
covar = list(CovSnd = cbind(ClassIntercept,ClassIsTeacher),
CovRec = cbind(ClassIsTeacher,ClassIsFemale)),
          ordinal = FALSE, hessian = TRUE)
Computing preliminary statistics
Fitting model
Obtaining goodness of fit statistics


> # This is preferred :
> classfit3$BIC classfit4$BIC
[1]1118.294


> # Next conversational norms + model 4
> classfit5 < rem.dyad(Class, n = 20,
          effects = c("CovSnd","CovRec","RRecSnd","RSndSnd",
"PSAB BA","PSAB AY","PSAB BY"),
          covar = list(CovSnd = cbind(ClassIntercept,ClassIsTeacher),
CovRec = cbind(ClassIsTeacher,ClassIsFemale)),
          ordinal = FALSE, hessian = TRUE)
Computing preliminary statistics
Fitting model
Obtaining goodness of fit statistics
```

> # Again an improvement :
> classfit4$*BIC* classfit5$*BIC*
[1]1699.716
> summary(classfit5)
Relational Event Model (Temporal Likelihood)

	Estimate	*Std.Err*	Z value	Pr$(>$ ‖z‖$)$
RRecSnd	2.429233	0.155365	15.6356	$<2.2e$ 16***
RSndSnd	0.986747	0.144667	6.8208	9.053e 12***
CovSnd.1	5.003434	0.090609	55.2201	$<2.2e$ 16***
CovSnd.2	1.253893	0.085160	14.7239	$<2.2e$ 16***
CovRec.1	0.722690	0.141950	5.0912	3.559e 07***
CovRec.2	0.047936	0.081325	0.5894	0.5556
PSAB *BA*	4.622128	0.137600	33.5910	$<2.2e$ 16***
PSAB *BY*	1.677591	0.164930	10.1715	$<2.2e$ 16***
PSAB *AY*	2.869968	0.103113	27.8332	$<2.2e$ 16***

Signif.codes : 0 '***' 0.001 '**' 0.01 '*' 0.05 '.' 0.1 ' '1
Null deviance : 5987.221 *on* 691 degrees of freedom
Residual deviance : 2803.315 *on* 683 degrees of freedom
     Chi square : 3183.906 *on* 8 degrees of freedom,
          asymptotic p value 0
AIC : 2821.315 AICC : 2821.58 *BIC* : 2862.158

We can see that adding recency effects to the covariate model results in a much improved fit by BIC. Moreover, there is again an improvement in BIC when conversational norms are added into the model. The summary of the results from model 5 also show that the remaining gender covariate effect falls out in the presence of the endogenous social dynamic effects. This hints at the possibility that what seemed at first glance to be a difference in the tendency to receive communication by gender was in fact a result of social dynamics (perhaps stemming from the fact that both instructors are male, with their inherent tendency to communicate more often amplified by local conversational norms). We can confirm that second the gender term is extraneous by evaluating whether a reduced model is preferred by BIC.

```
> classfit6 < rem.dyad(Class, n = 20,
            effects = c("CovSnd","CovRec","RRecSnd","RSndSnd",
"PSAB BA","PSAB AY","PSAB BY"),
            covar = list(CovSnd = cbind(ClassIntercept,ClassIsTeacher),
CovRec = ClassIsTeacher), ordinal = FALSE, hessian = TRUE)
Computing preliminary statistics
Fitting model
Obtaining goodness of fit statistics


> classfit5$AICC  classfit6$AICC
[1] 1.705912
```

And, as before, the reduced model is indeed preferred. We now have a relatively well-fitting relational event model specified by a combination of covariate and endogenous dynamic effects. At this point, we can turn to interpretation of fitted model parameters and model adequacy from our current vantage point.

4.5.3   Interpretation of a Fitted Model

It is often useful to consider the inter-event times predicted to be observed under various scenarios by a fitted relational event model. Recall that under the piecewise constant hazard assumption, event waiting times are conditionally exponentially distributed. This allows us to easily work out the consequences of various model effects for social dynamics, at least within the context of a particular scenario.

The most basic results to interpret from a fitted model are, of course, the coefficients themselves. In interpreting coefficient effects, recall that they act as logged hazard multipliers. Taking their log-inverse (i.e., exponentiating them), produces their hazard multiplier. For instance, the turn-taking participation-shift (p-shift) effect from model 6 has a coefficient value of 4.623682, which corresponds to an interpretation that response events have about 100 times the hazard of non-response events ($e^{4.623682} = 101.8684$). While this *appears* to be a substantial effect, the fact that an event has an unusually high hazard does not mean that it will necessarily occur. For instance, while a response of B to a communication from A has hazard that is about 100 times as great as the hazard of a non-B → A event all things constant, there are many more events of the latter type. In fact, there are 379 other events "competing" with the B → A event, and thus the chance that it will occur next is smaller than it may appear by simply taking the hazard multiplier at face value. This example shows that both relative rates and combinatorics (i.e., the number of possible ways that an event type may occur) govern the result and should temper respective interpretations.

What else can be done with the model coefficients from an interpretation perspective? One basic use of the model coefficients is to examine the expected inter-event times under specific scenarios and conditions. For instance, one may be interested in evaluating the predicted mean inter-event time when nothing else is happening. This is simply governed by the global pacing constant (i.e., the average rate that events transpire, or intercept) and the number of possible events. Or, one may want to know how long it takes for one actor to respond to another actor given an immediate event (or other such scenarios). Depending on the model, many of these "waiting time" effects can be evaluated from coefficients. To accomplish this using the exact time likelihood, some algebra comes in handy: $\dfrac{1}{m \times e^{\sum \times}}$ where m is the number of possible events under the scenario and λ is the vector of model parameters involving the scenario of interest. Here again, both the number of ways that an event type can occur (m') and the propensity of such events to occur (λ) both matter! In the following snippet, we evaluate such waiting times under different scenarios from model 6:

```
> # Mean inter  event time if nothing else going on....
> 1 / (20 * 19 * exp(classfit6$coef["CovSnd.1"]))
  CovSnd.1
0.3843285
```

```
> # Mean teacher  student time(again, if nothing else happened)
> 1 / (2 * 18 * exp(sum(classfit6$coef[c("CovSnd.1","CovSnd.2")])))
[1]1.153845
```

```
> # Sequential address by teacher w / out prior interaction,
given a prior teacher  student interaction, and assuming
nothing else happened
> 1 / (17 * exp(sum(classfit6$coef[c("CovSnd.1",,"CovSnd.2",,"PSAB AY")])))
[1]0.1384693
```

```
> # Teacher responding to a specific student, given an
immediate event
> 1 / (exp(sum(classfit6$coef[c("CovSnd.1",,,"CovSnd.2",,,"PSAB BA",,,"RRecSnd")])))
[1]0.03587346
```

```
> # Student responding to a specific teacher, given an
immediate event
> 1 / (exp(sum(classfit6$coef[c("CovSnd.1",,,"CovRec.1",,,"PSAB BA",,,"RRecSnd")])))
[1]0.2657102
```

Remember that our temporal units in the classroom dataset are increments of minutes: multiplying these values by 60 returns how many seconds (or fractions thereof) these predicted waiting times entail. Thus, if no other event were to intervene, a teacher would initiate communication with a student after a mean waiting time of approximately 70 seconds. Given an initial teacher→student communication and no other intervention, the same teacher will produce another speech act after an average of roughly 8 seconds—a rapid-fire lecture mode. Interestingly, we can also see that teachers are very quick to respond to student communications (a delay of just over 2 s, on average), while students take somewhat longer to respond to teachers (about 16 s). Such observations comport well with our general intuition regarding classroom functioning, and illustrate the types of quantitative information that can be gleaned from a REM fit.

4.5.4 Assessing Model Adequacy

We can assess model adequacy for exact time relational event models in much the same manner as we do for ordinal time models. The major difference is that we cannot here use a fixed null residual or guessing equivalent. However, we can still examine "surprise" based on the deviance residuals of fitted models. Despite not having a fixed null residual to evaluate against, we can still inspect the distribution of the deviance residuals. Ideally, we would like them to be small and clustered near zero. Figure 4.7 plots the histogram of the deviance residuals from model 6. The distribution is clearly more "lumpy" than that observed in Fig. 4.3 for the corresponding the WTC model, suggesting that the classroom dyamics are less well-predicted on average than were the radio communications.

```
> # Plot the histogram of the deviance residuals from model 6
> hist(classfit6$residuals)


> # How well do we predict the exact event?
> mean(apply(classfit6$predicted.match,,1,,all))
[1] 0.3299566


> # How well do we predict either the sender or receiver of an event?
> mean(apply(classfit6$predicted.match,,1,,any))
[1] 0.5166425


> # How well do we predict each part of the event?
> colMeans(classfit6$predicted.match)
    FromId        ToId
0.5050651   0.3415340
```

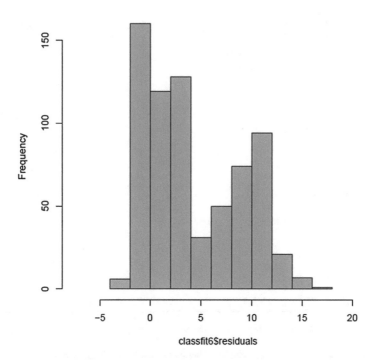

Fig. 4.7 Histogram of Deviance Residuals from Exact Time Model of McFarland's Classroom. Data

Evaluating how well the model predicts each event sheds additional light on these results.

On average, the model only predicts the event perfectly about 33 % of the time (still a remarkable performance, given the large number of possible events). We do a bit better with getting at least one part of the event right, correctly classifying the sender or receiver about 50 % of the time (and we do much better at classifying senders than receivers over all, on average). Moreover, inspection of the classification accuracy in Fig. 4.8 for this model shows substantial lag between the prediction threshold and fraction of the observed events covered by the model. By 25 % of the possible events transpiring, the model has only predicted 89 % of the observed events (compared with 98 % in the corresponding WTC case).

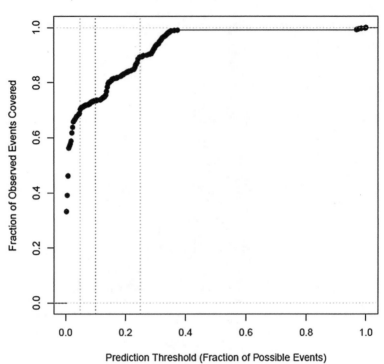

Fig. 4.8 Classification Accuracy of the Observed Ranks Under Model 6 with Prediction. Thresholds Indicated at 0.05, 0.1, and 0.25

```
> # Classification plot
> plot (ecdf (classfit6$observed.rank / (19 * 20) ), xlab = "Prediction
Threshold (Fraction of Possible Events)", ylab = "Fraction of
Observed Events Covered", main = "Classification Accuracy", xlim = c(0,1))
> abline (v = c(0.05,,0.1,,0.25), lty = 2)


> # a comparative look at the 25 th prediction threshold
> ecdf (classfit6$observed.rank / (19 * 20) )(.25)
[1] 0.8929088


> ecdf ( wtcfit6$observed.rank / (37 * 36) )(.25)
[1] 0.983368
```

Fig. 4.9 Time-Aggregated
'Surprising' Events
Network Under the Final
Relational Event. Model
of McFarland's Classroom
Data

So, comparatively, it looks that our exact time relational event model of the classroom data isn't performing as well as our ordinal time relational event model of the WTC data. We may be missing some important aspect of the relational event process in our model of the classroom conversation. We can again examine the model "surprise" superimposed on the time-aggregated network for clues about what may be going on. Here, because we lack a null residual, we'll define surprising events as those for which the observed event is not in the top 5 % of those predicted.

```
> #rank > 19 corresponds to the 5 \ %cut  off here
> surprising <  as.sociomatrix.eventlist(Class[classfit6$observed.rank > 19,],20)

> #Plot the resulting surprising network
> gplot(surprising, edge.lwd = surprising ^ .75,
        arrowhead.cex = log(as.edgelist.sna(surprising)
[,3]) + .25, vertex.col = ifelse(WTCPoliceIsICR,,"black",,"gray"),
        vertex.cex = 1.25,
vertex.sides = ifelse(WTCPoliceIsICR,,4,,100), displayisolates = FALSE)
```

The visualization in Fig. 4.9 gives us more of a clue about what we're missing. Specifically, the presence of five distinct clusters represent the occurrence of various side discussions that are not well-captured by the current model. This could be due to the fact that things like P-shift effects fail to capture simultaneous side-conversations (each of which may have its own set of turn-taking patterns), or to a lack of covariates that capture the enhanced propensity of subgroup members to address each other (such as students being in the same school club together). Further elaboration could be helpful here. On the other hand, we seem to be doing reasonably well at capturing the main line of discussion within the classroom, particularly vis-a-vis the instructors. Whether or not this is adequate depends on the purpose to which the model is to be put; as always, adequacy must be considered in light of specific scientific goals.

4.6   Conclusion

A wide range of interaction processes—from radio communications to dominance contests— can be fruitfully studied within the relational event paradigm. While arising as the short duration limit of the dynamic network regime, the relational event regime has its own distinct properties and requires distinct treatment. In particular, relational event dynamics are fundamentally about *sequential relational structure*, rather than the *simultaneous relational structure* that is the dominant concern within social network analysis. In this and many other respects, theory and analysis of relational event dynamics owes as much to fields such as conversation analysis, event history analysis, and agent-based modeling as to conventional network analysis. Relational event models are still fundamentally structural, however, and we stress that the approaches are complementary. Indeed, where exact (or exactly ordered) data is available on relationship start and stop times, it is possible to model dynamic networks via a REM process whose events involve the creation and termination of edges. Taking such a process to be fully latent—with only the state of the currently active edges observed at a small number of distinct points in time—leads one to a model family that is essentially similar to the framework of Snijders (2001). Likewise, temporally extensive relationships are often important covariates for relational event processes, allowing one to directly assess the impact of ongoing ties on social microdynamics.

Although we have focused here on some of the most basic types of REMs, more complex cases are also possible. As noted, REMs for "egocentric" event data (Marcum & Butts, 2015) can be powerful tools for modeling the responses of individuals to their local social environments, and are well-suited to the analysis of complex event series (with many event types) punctuated by exogenous events. Hierarchical extensions to REMs (DuBois et al., 2013b) allow for pooling of information across multiple event sequences while still allowing the dynamics of each sequence to differ from the others; this is particularly useful when studying many small groups, and/or when attempting to estimate covariate effects for attributes whose prevalence varies greatly from group to group. Endowing REMs with latent structure also holds a host of opportunities, including the ability to infer latent interaction roles directly from behavioral data (DuBois, Butts, & Smyth, 2013a). Given the breadth and flexibility of the approach, the prospects are good for many more developments in this area. We close with the important reminder that no representation is fit for all purposes, nor is it intended to be. Many relational analysis problems involve the modeling of ongoing relationships, and are better viewed through the lenses of static or dynamic network analysis. Where one's focus is on micro-interaction or other processes involving discrete behaviors whose implications cascade forward through time, however, the relational event paradigm offers a powerful and statistically grounded alternative.

References

Almquist, Z. W., & Butts, C. T. (2014). Logistic network regression for scalable analysis of networks with joint edge/vertex dynamics. *Sociological Methodology, 44*(1), 273–321.

Bender-deMoll, S., & McFarland, D. (2006). The art and science of dynamic network visualization. *Journal of Social Structure, 7*(2).

Blossfeld, H. P., & Rohwer, G. (1995). Techniques of Event History Modeling: New Approaches to Causal Analysis. Lawrence Erlbaum and Associates, Mahwah, NJ.

Burt, R. S. (1992). *Structural holes: The social structure of competition.* Cambridge, MA: Harvard University Press.

Butts, C. T. (2008). A relational event framework for social action. *Sociological Methodology, 38*(1), 155–200.

Butts, C. T. (2009). Revisiting the foundations of network analysis. *Science, 325,* 414–416.

Butts, C. T. (2010). *Relevent: Relational event models.* R package version 1.0.

Butts, C. T., Petrescu-Prahova, M., & Remy Cross, B. (2007). Responder communication networks in the world trade center disaster: Implications for modeling of communication within emergency settings. *Mathematical Sociology, 31*(2), 121–147.

Centola, D., & Macy, M. (2007). Complex contagions and the weakness of long ties. *American Journal of Sociology, 113*(3), 702–734.

DuBois, C., Butts, C., & Smyth, P. (2013a). Stochastic blockmodeling of relational event dynamics. In *Proceedings of the Sixteenth International Conference on Artificial Intelligence and Statistics,* 238–246.

DuBois, C., Butts, C. T., McFarland, D., & Smyth, P. (2013b). Hierarchical models for relational event sequences. *Journal of Mathematical Psychology, 57*(6), 297–309.

Freidkin, N. (1998). *A structural theory of social influence.* Cambridge: Cambridge University Press.

Gibson, D. R. (2003). Participation shifts: Order and differentiation in group conversation. *Social Forces, 81*(4), 1335–1381.

Granovetter, M. (1973). The strength of weak ties. *American Journal of Sociology, 78*(6), 1369–1380.

Heider, F. (1946). Attitudes and cognitive organization. *Journal of Psychology, 21,* 107–112.

Koskinen, J. H., & Snijders, T. A. (2007). Bayesian inference for dynamic social network data. *Journal of Statistical Planning and Inference, 137*(12), 3930–3938.

Krivitsky, P. N., & Handcock, M. S. (2014). A separable model for dynamic networks. *Journal of the Rotal Statistical Society, Series B, 76*(1), 29–46.

Lakon, C. M., Hipp, J. R., Wang, C., Butts, C. T., & Jose, R. (2015). Simulating dynamic network models and adolescent smoking: The impact of varying peer influence and peer selection. *American Journal of Public Health, 105*(12), 2438–2448.

Leenders, R., Contractor, N. S., & DeChurch, L. A. (2015). Once upon a time: Understanding team dynamics as relational event networks. *Organizational Psychology Review., 6*(1), 92–115.

Liang, H. (2014). The organizational principles of online political discussion: A relational event stream model for analysis of web forum deliberation. *Human Communication Research, 40*(4), 483–507.

Marcum, C. S., & Butts, C. T. (2015). Creating sequence statistics for egocentric relational events models using informr. *Journal of Statistical Software*, *64*(5), 1–34.

Mayer, K. U., & Tuma, N. B. (1990). *Event history analysis in life course research*. Madison, WI: University of Wisconsin Press.

McFarland, D. (2001). Student resistance: How the formal and informal organization of classrooms facilitate everyday forms of student defiance. *American Journal of Sociology*, *107*(3), 612–678.

Mills, M. (2011). *Introducing survival and event history analysis*. Thousand Oaks, CA: Sage.

Morris, M., Goodreau, S., & Moody, J. (2007). Sexual networks, concurrency, and STD/HIV. In K. K. Holmes, P. F. Sparling, W. E. Stamm, P. Piot, J. N. Wasserheit, & L. Corey (Eds.), *Sexually transmitted diseases* (pp. 109–126). New York: McGraw-Hill.

Patison, K., Quintane, E., Swain, D., Robins, G. L., & Pattison, P. (2015). Time is of the essence: An application of a relational event model for animal social networks. *Behavioral Ecology and Sociobiology*, *69*(5), 841–855.

Petrescu-Prahova, M., & Butts, C. T. (2008). Emergent coordinators in the World Trade Center Disaster. *International Journal of Mass Emergencies and Disasters*, *26*(3), 133–168.

Price, D. (1976). A general theory of bibliometric and other cumulative advantage processes. *Journal of the American society for Information Science*, *27*(5), 292–306.

Rapoport, A. (1949). Outline of a probabilistic approach to animal sociology. *Bulletin of Mathematical Biophysics*, *11*, 183–196.

Robins, G. L., & Pattison, P. (2001). Random graph models for temporal processes in social networks. *Mathematical Sociology*, *25*, 5–41.

Sampson, S. (1969). *Crisis in a cloister*. Doctoral Dissertation: Cornell University.

Snijders, T. A. (1996). Stochastic actor-oriented models for network change. *Mathematical Sociology*, *23*, 149–172.

Snijders, T. A. B. (2001). The statistical evaluation of social network dynamics. *Sociological Methodology*, *31*, 361–395.

Tranmer, M., Marcum, C. S., Morton, F. B., Croft, D. P., & de Kort, S. R. (2015). Using the relational event model (rem) to investigate the temporal dynamics of animal social networks. *Animal Behaviour*, *101*, 99–105.

Wang, C., Hipp, J. R., Butts, C. T., Jose, R., & Lakon, C. M. (2016). Coevolution of adolescent friendship networks and smoking and drinking behaviors with consideration of parental influence. *Psychology of Addictive Behaviors*, *30*(3), 312–324.

Wasserman, S., & Faust, K. (1994). *Social network analysis: Methods and applications*. Cambridge: Cambridge University Press, Cambridge.

Wasserman, S., & Robins, G. L. (2005). An introduction to random graphs, dependence graphs, and p_. In P. J. Carrington, J. Scott, & S. Wasserman (Eds.), *Models and methods in social network analysis* (pp. 192–214). Cambridge: Cambridge University Press.

Chapter 5
Text Mining Tutorial

Natalie J. Lambert

5.1 Introduction

The world we live in is generating text at an unprecedented rate. Consider how much new text is created by emails, newspapers, blogs, and social media websites every day, and it quickly becomes clear that analysis of group behaviors can become challenging due to the large amount and variety of textual data generated from group members' interactions. Text mining is one strategy for analyzing textual data archives that are too large to read and code by hand, and for identifying patterns within textual data that cannot be easily found using other methods. Text mining as a method can be used to conduct basic exploration of textual data, or can be used in combination with other methods like machine learning to predict group members' future behaviors. This tutorial introduces text mining by outlining two basic methods for data exploration: generation of a concept list and generation of a semantic network. Learning the steps it takes to prepare, import, and analyze textual data for these simple procedures is enough to get started analyzing your own datasets. This tutorial is only a glimpse of the text mining method, however, and new text mining programs and algorithms are continually being developed. Readers interested in learning more about text mining should take formal courses or explore the many text mining packages available in programming languages like R and Python.

Most fundamentally, text mining is a methodology used to extract information, classify data, and identify patterns within textual datasets. It is even more accurate to say that text mining is a collection of methodologies because just as there can be many patterns within any one collection of text, there are many ways to identify these patterns using text mining. Historically, text mining was used to search

N.J. Lambert (✉)
Brian Lamb School of Communication, Purdue University, West Lafayette, IN, USA
e-mail: njlambert@purdue.edu

© Springer International Publishing AG 2017
A. Pilny, M.S. Poole (eds.), *Group Processes*, Computational Social Sciences,
DOI 10.1007/978-3-319-48941-4_5

computer documents in order to identify which documents contained a word or words of interest, and to extract specific information from documents (Fan, Wallace, Rich, & Zhang, 2006). Early electronic card catalogs in libraries utilized text mining to tag and index catalogue holdings (Miner, 2012), and text mining has been used to automatically generate research article abstracts from the content of articles since the 1950s (Luhn, 1958). Text mining is used today by businesses and researchers for a multitude of purposes such as analyzing news stories in order to understand the public's perception of health topics like AIDS (Caputo, Giacchetta, & Langher, 2016), to extract trends in consumer opinions from product reviews posted online (Dasgupta & Sengupta, 2016), and to manage information overload in research fields like biomedical research (Cohen & Hersh, 2005).

There are many situations where other methodologies cannot provide the type of information about a textual dataset that text mining can offer. A researcher with 60 hours of audio recordings of focus group interviews is faced with around 1,800 pages of transcriptions. Hand coding of such data for a factor of interest usually requires multiple readings of the text by several researchers, and such large textual datasets are often a daunting barrier to analysis even when they offer significant benefits like coverage of a greater variety of research subject demographics and backgrounds. Text mining can search through these large datasets for evidence of a factor of interest in seconds as opposed to the many hours it would take to manually search all of the transcriptions.

Another benefit of text mining is its ability to perform data-driven discovery. Data-driven discovery is the process of looking for patterns within datasets without pre-conceived hypotheses regarding what the researcher expects to find. Using the traditional scientific method, the researcher with the large archive of focus group transcriptions would have analyzed the data in order to answer a specific hypothesis such as, "Organizational groups that utilize a cooperative approach to conflict will attain higher productivity ratings than organizational groups that utilize a competitive approach to conflict." The researcher would likely answer this hypothesis by focusing on instances of conflict within the transcriptions, using a method like structural equation modeling to evaluate whether there is a relationship between group conflict style and the groups' productivity. Data-driven discovery conducted using text mining allows the researcher to broaden his or her focus to *anything* within the transcriptions that is significant to the conversation generated during the focus groups. Topic modeling or cluster analysis of a semantic network generated from the transcriptions could reveal a number of frequently-occurring topics like wage gaps or understaffing that a hypothesis-driven approach not focused on these topics would be unlikely to identify. Text mining can also be used in combination with other methods to double-check whether any frequently-occurring themes or words are present within the data that were not recognized by other forms of analysis. Text mining should not, however, be considered in any way superior to traditional research methods—it simply offers a new approach to examining textual data and is especially useful for managing data overload.

5.2 Overview of Text Mining

There are many analyses that can be performed using text mining, but the way in which the method operates is similar for most text mining procedures. During a text mining procedure, an algorithm built into the software contains a set of instructions for how to examine the text data and what to make note of. For example, during the first phase of analysis, called preprocessing (described in more detail below), the algorithm for the procedure called "stop word removal" tells the software to look word by word through the text data for all the words on a "stop word list," a list of words the researcher wants to exclude from analysis. The software "reads" through the entire dataset one word at a time, comparing each word in the dataset to the words on the stop word list, removing all words from the dataset that match a word found on the list. Another common procedure in text mining is the generation of a concept list, which is an inventory of all of the words in a dataset along with a count of how frequently each word appears in a dataset. The algorithm that creates the concept list also passes through the text word by word, adding new words it encounters to the concept list and adding a count to a word's tally number each time it reencounters the same word in the text. There are many more sophisticated ways that text mining algorithms draw information from a text archive than those just described, but the basic principle is that an algorithm contains a set of instructions for how the software should read and keep track of information found within the text. A full text mining analysis almost always involves running multiple procedures in a particular order in order to extract the information a researcher is interested in from the text.

As the reader likely can imagine, text mining as a method has some very specific assumptions built into it. The biggest assumption is certainly that individual words can have meaning even when they are far removed from their original context. A concept list, for example, counts the total number of times each word in a dataset appears within the text without taking the specific context where each word was used into account. The word "hate" means something very different when someone says "I hate my job" and "I'd hate to lose my job," but a standard concept list cannot tell you that. Data scientists are building algorithms and text mining approaches that can take the context of all words into account (see Lexalytics, 2015), but for scholars new to text mining it is important to remember that words spelled the same but with different meanings can be counted as the same concept. Another common assumption of text mining is that frequently-occurring words within a text archive are more significant than infrequently occurring words. This may indeed be the case, as it is in the tutorial example, or a word could simply occur frequently because it is a commonly used word for a certain language or context. There are also cases where word frequency is completely unimportant for understanding a particular dataset. It is therefore the analyst's responsibility to think through algorithms' built-in assumptions when performing text mining.

A third assumption of many text mining algorithms is that words that occur near each other in a text archive are related in some way. This chapter will demonstrate how to generate a semantic network, which is a group of words existing within a text archive that have been found to share some sort of relationship in common. According to many text mining algorithms, what these words usually have in common is proximity. Text mining tools commonly assign two words to the same meaning group when they both occur within a certain distance of each other within the text. It seems safe to assume that words that occur within the same sentence or paragraph are related, but if we look at the "I hate my job" and "I'd hate to lose my job" example again, it possible to see how words that occur near one another in textual data can be related but also have context-specific meanings that can be overlooked by algorithms only interested in words' relative positioning.

These examples are not meant to foster mistrust in text mining, but rather for the reader to gain an understanding of what the method can and cannot do. Text mining can provide a researcher with valuable information about his or her data such as which people, organizations, and places feature prominently within it, analyzed through a procedure called entity detection. Text mining can give a data analyst a sense of the emotion being expressed during conversations through a procedure called sentiment analysis. The text mining method can also map out dominant conversations taking place within communication datasets, showing where there is overlap between conversation topics. Or, text mining can be used to identify important phrases or patterns within business reports in order to expose reoccurring problems (Choudhary, Oluikpe, Harding, & Carrillo, 2009) and to detect public health rumors online (Collier et al., 2008). Google Book's Ngram viewer (http://books.google.com/ngrams) is an example of how simply tracking word frequencies over time can result in a sense of the rise and fall of the public interest in different topics. Figure 5.1 visualizes a comparison of the frequency of the appearance of the words "war" and "peace" over time in Google's large book archive. Note the rise in the term "war" following the first and second world wars. The graph also indicates that although books contained the word "war" more frequently than "peace," the appearance of "war" and "peace" followed very similar patterns.

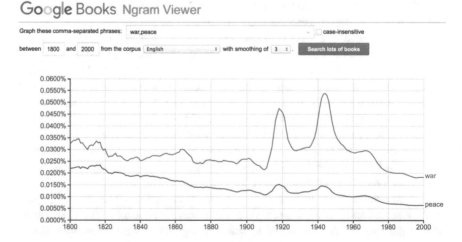

Fig. 5.1 Google Books Ngram viewer graph of the words "war" and "peace"

The result of text mining analysis is a summary of a pattern identified by the procedure that was run. The form this patterns takes can vary quite a bit, from a simple concept list to a very complex semantic network map or a new document file containing extracted data that fits the parameters the algorithm was designed to find, such as a subconversation. The results of text mining reveal something about what words or sections of a text archive are meaningful either because they occur frequently, are closely related to other words within the text, or because they fit some other parameter set by the researcher. Text mining analysis is not complete, however, until the analyst has reexamined the results back within the context of the data in order to interpret the meaning of the pattern. While the patterns text mining can reveal often seem self-explanatory, a deeper understanding of the data is only gained by assessing why certain words were found to be related and not others, and what this means for the group being studied.

5.3 Text Mining Tutorial

There are many methods that can be used to conduct exploratory text mining. This tutorial covers basic preprocessing steps as well as the generation of a concept list and semantic network. These text mining techniques will be demonstrated using AutoMap (Carley, 2001), a text mining tool developed by the CASOS Group at Carnegie Mellon University. (See Carley, Columbus, Bigrigg, Diesner, and Kunkel (2010) for a tutorial.) There are dozens of text mining tools, each with their individual benefits and suitable for different analyses and types of datasets. Tools like AutoMap that have graphical user interfaces are excellent for beginners interested in exploratory text mining. Once an analyst is comfortable with basic text mining, however, he or she will likely need to learn some programming skills in order to perform advanced procedures customized to his or her particular dataset.

The overall process for conducting text mining is: (1) data collection, (2) data preparation, (3) pre-processing, (4) analysis, and (5) interpretation. This tutorial will take you through each step of the method by describing an analysis conducted for a research project which examined small groups of emergency medical physicians as they drew on their professional expertise during medical consultations in order to develop patient treatment plans (Lammers, Lambert, Abendschein, Reynolds-Tylus, & Varava, 2016).

5.3.1 Data Collection

The sample text corpus used in this tutorial was collected during a study of medical consultations taking place in the emergency department of a hospital. The emergency department was staffed by about two dozen full-time physicians, including doctors, physicians' assistants, nurse practitioners, and medical residents. The team of researchers was permitted to observe physicians' conversations with one another in their shared office space away from patients. The researchers transcribed by hand, as verbatim as was possible, the conversations between physicians related to patients' care in the emergency department. They also noted which physician initiated each conversation and which physicians participated in each conversation. The data collection totaled 90 h of observations, which resulted in 159 pages of field notes and a text corpus of medical consultations containing 19,868 words. The following is a hypothetical example of a typical medical consultation observed by the research team, created in order to preserve participants' privacy:

Doctor: What's going on with room 23?
Resident: He's a 42-year-old man, diabetic. Complaining of pain in abdomen and side. No fever, white count is normal.
Doctor: Possible kidney stone. Any pain medicine prescribed? You can give him morphine.
Resident: Sounds good.

The research team was interested in studying medical consultations because existing research had shown that communication problems between physicians can result in treatment errors, especially during patient handoffs (Maughan, Lei, & Cydulka, 2011). Medical professionals had also called for a better understanding of medical consultations beyond exploratory studies offering models and taxonomies of medical consultations (Kessler et al., 2011). Little was known about what a medical consultation looks like or what topics or problems physicians encounter during consultation, and so that is what the research team set out to learn by collecting and analyzing empirical observations of medical consultations. Their goal was to distinguish between different types or topics of medical consultations in order to better understand how medical professionals enact expertise. Text mining was a useful method for this research project because the data collected by the team was unstructured textual data, meaning the data was in its naturally occurring form and not classified or organized into a database. The researchers knew very little about the data since no one had ever looked at the topics surrounding medical consultations before. An exploratory method that could look for patterns within the textual data was therefore the best fit, and that is what text mining is designed to do.

5.3.2   Data Preparation

After collecting a textual dataset, the next step of the text mining method is to prepare the data for analysis. Data preparation involves removing all data items from a text archive (often called a text corpus) except for the text of interest, and converting the data into a format that the software can import and read. In the case of the example research team, once they decided that they wanted to analyze medical consultations between all the physician role types, they removed the role labels from the text corpus (i.e., Doctor, PA, etc.) so that only the transcribed medical consultations remained. The next step was to copy all the text transcriptions and paste them into Notepad. When using AutoMap and many other text mining tools, the file

Fig. 5.2 Data and output folder creation

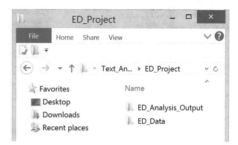

Fig. 5.3 The AutoMap home screen

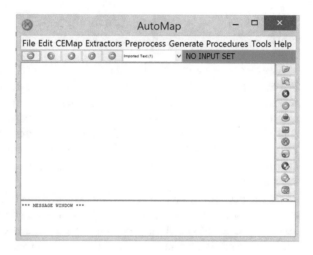

extension of the data file must be ".txt", because the data must be contained within in a plain text file in order for the software to be able to read it. Other text editors can be used instead of Notepad as long as they do not preserve file formatting and can generate plain text files. The research team data analyst saved the plain text file containing the medical consultation data within a new folder, and did not place anything else in the folder. If a dataset is comprised of multiple text files, the analyst should place all of the text files he or she wants to analyze simultaneously within this folder. AutoMap will import all files within the folder as one dataset. The research team data analyst next created another empty folder where the data analysis output would be stored. The dataset and analysis output folders can be seen in Fig. 5.2.

After properly formatting and storing the medical consultation data, the data analyst imported the data file into AutoMap. To do this, the analyst began at the AutoMap home screen (Fig. 5.3) and imported the data file by clicking on *File— Import Text Files*. The next step was to click once to highlight the file folder containing the data, and then click *Select* (Fig. 5.4). The data analyst used the preselected settings for text encoding and text direction and pressed *Enter*.

Fig. 5.4 Importing a text
corpus into AutoMap

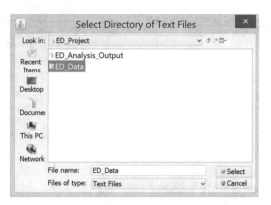

The text contained within the data file was imported into AutoMap and displayed in the text display pane. Due to privacy agreements with the example study's research subjects, this tutorial cannot show the transcription of the medical consultation dataset. As an alternative, the full script of Shakespeare's *Romeo and Juliet* (Fig. 5.5) has been imported into AutoMap using the previously described steps. This tutorial will use *Romeo and Juliet* as dummy data to demonstrate the next step, data preprocessing, and then return to the medical consultation transcriptions to show the results of a real data analysis.

As can been seen in Fig. 5.5, the all-caps indicators of the act, scene, and characters are included in the imported data file. This was done in order to learn more about the main features of the play, and because the analyst decided in advance that characters were important features of the play and therefore should be included in data analysis. If the analyst was instead interested in analyzing the dialogue of the play and wanted to compare and contrast different characters' dialogue, her or she would have collected each character's lines into separate plain text files and removed all-caps text and any other non-dialogue text from the files. Each file would be analyzed separately and comparisons made of the individual analysis results for each character. Data preparation is a very important part of the text mining method because during this step the analyst must make choices about what selections of a larger text corpus to include in the analysis. Every text corpus contains different characteristics that must be taken into consideration when making decisions about how to best prepare data to answer a specific research question.

5.3.3 Preprocessing

Once the dataset has been formatted and imported into AutoMap, the next step is preprocessing of the data. Preprocessing is a term used to describe the cleaning up and standardizing of textual data prior to analysis. Two common types of preprocessing are stop word removal and stemming. Stop words are any words that would

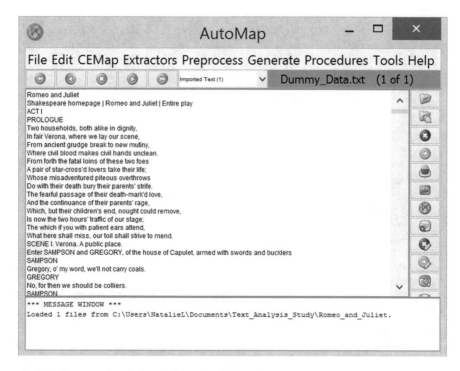

Fig. 5.5 *Romeo and Juliet* imported into AutoMap

interfere in the software's ability to identify meaningful patterns within the data. These are usually high frequency words that do not have a lot of significance for most datasets such as articles, conjunctions, pronouns, number words, contractions, simple verbs, and prepositions. Many text analyzers have built-in stop word lists (also called "delete" lists), but a researcher can also create his or her own by making a list of words that are known to be frequent within a dataset but do not add value to the analysis.

To perform stop word removal on the *Romeo and Juliet* dataset within AutoMap, the data analyst clicked on ***Preprocess—Text Refinement—Apply Delete List***, and clicked ***Confirm***. She used the standard AutoMap delete list (which contains the most frequently-occurring words within the English language: a, an, and, as, at, but, for, he, her, hers, him, his, etc.), although she could have edited the delete list within AutoMap to create a custom list. The next step was to select *Rhetorical* as the type of delete processing because this setting inserts a placeholder, *xxx,* into the data so that the analyst can see which words were removed due to this procedure. Rhetorical delete processing also preserves the distance between words so that two words are not considered closer together after the words on the delete list that exist between them are removed. The analyst clicked ***OK***, and the results of delete list application can be seen in Fig. 5.6. Note that you can see the list of procedures that have been performed on the data so far in the Message Window.

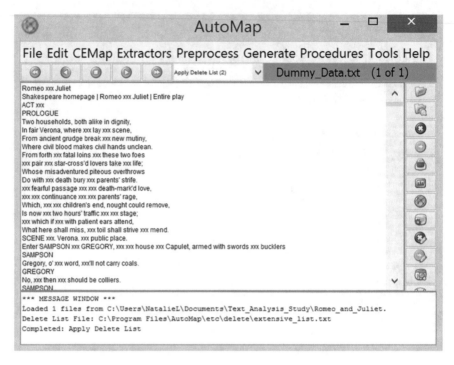

Fig. 5.6 Delete list applied to *Romeo and Juliet* dataset

Further preprocessing can be done by using the ***Preprocess—Text Preparation*** functions such as ***Remove Numbers as Words or Remove All Noise Words***. The analyst chose to apply ***Remove All Noise Words*** to the *Romeo and Juliet* dataset because this procedure removes pronouns, verbs, possessives, number words and other words types that researchers often find beneficial to remove from their textual data before analysis. The amount and type of preprocessing that should be performed depends on the dataset and what the researcher wishes to learn from it. For example, in some datasets pronouns could be important indicators of personal identification, and inclusion of all verbs might be important for analysis of storytelling or for identifying time phases. It is up to the researcher to evaluate the benefits and impact of specific preprocessing techniques on a particular dataset. Figure 5.7 shows the dataset after all noise words were removed.

The second preprocessing technique, stemming, involves identifying the root of a word and then standardizing all the various endings that come after a root in order to avoid separate counts of a word that has different forms but the same meaning. For example, the words "live," "lived," and "lives" would all be considered unique words by a text analyzer unless the analyst performed preprocessing like stemming that can reconcile these differences within the dataset. After performing stemming, the root of these words, "live", would take the place of all other forms of the word within the dataset. The analyst applied stemming to the *Romeo and Juliet* dataset in

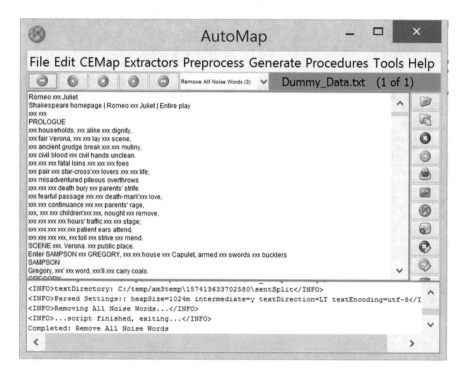

Fig. 5.7 *Romeo and Juliet* dataset after all noise words removed

order to demonstrate this procedure. In AutoMap stemming is conducted by clicking on *Preprocess—Text Refinement—Apply Stemming*. The analyst used the default *K-stemmer*, clicked *OK*, chose the default option to include capitalized words in stemming, and clicked *OK* again. The results (Fig. 5.8) show that verbs have been converted to their root form, so that "lay" became "lie." Plural nouns like "ears" were converted to singular nouns, and all words not in their root form were brought to their root form. It is now much less likely that words with the same meaning will be analyzed separately because of grammar or conjugation factors.

5.3.4 Data Analysis

Text Corpus Statistics. Now that the *Romeo and Juliet* dataset has been preprocessed, the simplest type of exploratory analysis that can be done is generation of a concept list. As mentioned earlier, a concept list is a inventory list of the words that appear within a text corpus along with a count of each word's frequency and other attribute information. The analyst generated a concept list for the *Romeo and Juliet* dataset by clicking on *Generate—Concept List—Concept List (Per Text)*. AutoMap's request to "Select Directory for Concept Lists" asks the analyst to select

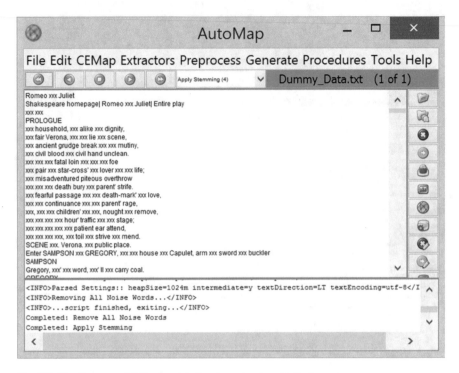

Fig. 5.8 The *Romeo and Juliet* dataset after stemming was applied

an output folder where he or she wants the results of the analysis to be stored. The analyst should only click once on the output folder to highlight it, then click **Select**. The next window allows the analyst to specify some concept list generation parameters. For this example the analyst used the default parameters and then clicked **Confirm**. AutoMap gives the option to open the concept list in its built-in viewer window, but the user can also navigate to the output folder on his or her computer where a new folder, *Concept List1*, has been created to store the concept list. The concept list is created as a Microsoft Excel file, which makes it convenient to sort the list according to the frequency that a word appears in the dataset, or according to any other attribute assigned by the researcher. The analyst opened the file in Excel and then sorted the list by frequency, as seen in Table 5.1. The concept list shows each word within the corpus, a count of how frequently it occurred within the corpus, and a relative frequency score compared to the concept that occurred most often in the corpus.

Even though the concept list is a very simple text mining method, it does reveal some meaningful information about the data, especially for people who have never read *Romeo and Juliet* or seen the play performed. The concept list can be interpreted as evidence that a large amount of the text is devoted to a love story in which two characters, Romeo and Juliet, factor highly. The list indicates that night may be an important time or setting of the play, and that a nurse, friar, and people named Mercutio, Benvolio, and Laurence are important characters. The word "death"

Table 5.1 Concept list generated from the *Romeo and Juliet* dataset

	A	B	C
1	Concept	Frequency	relative_frequency
2	ROMEO	180	1
3	love	155	0.8611111
4	thy	150	0.8333333
5	thee	138	0.76666665
6	JULIET	134	0.74444443
7	Romeo	130	0.7222222
8	CAPULET	119	0.6611111
9	Nurse	114	0.6333333
10	ll	91	0.50555557
11	BENVOLIO	74	0.41111112
12	night	73	0.40555555
13	Enter	72	0.4
14	FRIAR	70	0.3888889
15	MERCUTIO	69	0.38333333
16	man	69	0.38333333
17	LAURENCE	65	0.3611111
18	good	65	0.3611111
19	death	64	0.35555556
20	LADY	62	0.34444445

appears in the text relatively frequently, and so the analyst might assume that one or several characters die—this might therefore be a romantic tragedy. The concept list provides only a very basic understanding of the play that is divorced from its prose and plot, but perhaps through this example the reader can now visualize how text mining can aid researchers in extracting meaningful information from text corpuses much larger than a play that would otherwise take weeks to read through and summarize.

The concept list also points out where preprocessing improvements are necessary. The list shows that ROMEO and Romeo were counted separately by AutoMap. This result can be considered useful in that it distinguishes between the play formatting made in all caps and the verbal references to Romeo that appeared during the play, but it could also be considered an error if the analyst's goal was to count all mentions of Romeo together. "Thy" and "thee" show up at the top of the list because the delete list was created with modern language in mind. These pronouns should be added to the delete list and preprocessing rerun. The word "ll" needs to be investigated since it may be a result of stemming or could be part of an archaic word in the text that could not be properly preprocessed. *As the reader can see, text mining analyses must often be repeated multiple times in order to refine preprocessing to suit the nuances of each dataset.* Conversely, in some cases the analyst may want to do very little preprocessing in order to preserve all variation within the data for analysis. This was the case for the medical consultation dataset. Due to its smaller

Window Position 1	xxx household, xxx alike xxx dignity, xxx fair Verona, xxx xxx lie xxx scene,
Window Position 2	xxx ancient grudge break xxx xxx mutiny, xxx civil blood xxx civil hand unclean.
Window Position 3	xxx xxx xxx fatal loins xxx xxx xxx foe xxx pair xxx star-cross'xxx lovers xxx xxx life;
Window Position 4	xxx misadventured piteous overthrow xxx xxx xxx death bury xxx parent' strife.
Window Position 5	xxx fearful passage xxx xxx death-mark' love, xxx xxx continuance xxx xxx parent' rage,

Fig. 5.9 Illustration created to demonstrate how AutoMap creates "windows" to extract word pairs during semantic network generation

size, stemming made it impossible to detect the nuances of conversations surrounding similar medical consultation topics. As a result the analyst only performed stop word removal and removal of numbers as words when preprocessing the medical consultation dataset.

Semantic Network Analysis. This tutorial now returns to the medical consultation dataset in order to demonstrate how to generate a semantic network from a text corpus. Generation of a concept list using the medical consultation dataset revealed that "pain" was a frequently occurring word within the text corpus as was "goodbye." The research team wanted to know more about the context of these and other frequently occurring words, and so the team's data analyst constructed a co-occurrence semantic network from the data.

Co-occurrence semantic networks are based upon two key notions: (1) the idea that words that exist close to each other within a textual dataset are likely related in some way, and (2) that the meaning of a text corpus can be analyzed by constructing a network that represents all of the relationships between words in a dataset simultaneously. Take for example the sentence, "The patient complains of pain in his abdomen." Stop word removal would leave us with: "patient complains pain abdomen." Because these words occur near each other (within the same sentence), AutoMap makes note of their proximal relationship. The specific way in which the software does this is as follows. AutoMap creates a "window," the size of which is specified by the analyst (for example, two sentences or a paragraph in size) and then moves the window through the data, looking at the text that fits within the window and keeping track of the words that appear within the same window. (Figure 5.9 is an illustration of how a two-sentence window would move through the *Romeo and Juliet* dataset.) By repeating this procedure throughout the data, AutoMap collects a count of how many times a pair of words like "patient" and "pain" co-occur with one another within the same window. The resulting list of word pairs can be visualized as a network that connects all the pairs to one another so that if "pain" and "patient" co-occur frequently, and "chest" and "pain" co-occur frequently, one branch of the network will look like this: patient—pain—chest. In a network visualization, the lines that connect the words, called edges, can be used to represent how many times the same pair of words co-occurs within the dataset by thickening the width of the line to represent a greater frequency of co-occurrence.

The first step in constructing a co-occurrence semantic network is to click on **Generate—Semantic Network—Semantic (Co-reference List)**. The analyst again clicked once on the output folder to select it, then clicked **Next**. The network parameters window allows the analyst to make several decisions about how to generate the network. Directionality refers to whether the edge between two words represents a unidirectional (one-way flow or relationship) or bidirectional relationship (two-way flow or mutual relationship). For the medical consultation network, the analyst chose to setup the network as having bidirectional relationships because the research team wanted to discover the relationships between words within the medical consultation conversations without putting a word order constraint on the network. For their project, "doctor-patient" and "patient-doctor" could be counted as the same word pair because word order would not change which concepts were related topically to one another. Word order had the potential to cause variation in the meaning of these topics, but that was something the analyst was aware she would need to evaluate. Analysis of the network with no word order constraints was her team's best option for a first round of data analysis. Therefore, if two words co-occurred within the same window, the software noted their mutual, proximal relationship. If the team had been interested in identifying frequently-occurring phrases, they would have needed to preserve the order of words within each sentence and would have chosen instead to generate a unidirectional network. The analyst selected the window size as a two sentence window because of the small size of the text corpus, left the other parameters at their default values, and clicked **Confirm**. This analysis generates a folder, **SemanticList1**, within the output folder. The output itself is an Excel file containing two columns that represent pairs of words extracted using the described windowing method, along with a column that is a record of how frequently each pair of words occurred within the text corpus (Fig. 5.10).

Fig. 5.10 The semantic word pair list resulting from semantic network generation from the medical consultation dataset

	A	B	C
1	source_id	target_id	frequency
2	chest	pain	18
3	sounds	good	17
4	chest	x-ray	13
5	stress	test	9
6	emergency	department	8
7	ago	days	6
8	ago	weeks	6
9	bowel	obstruction	6
10	care	primary	6
11	huh	uh	6
12	feeling	better	5
13	ct	scan	5
14	abdominal	pain	4
15	blood	pressure	4
16	blood	white	4

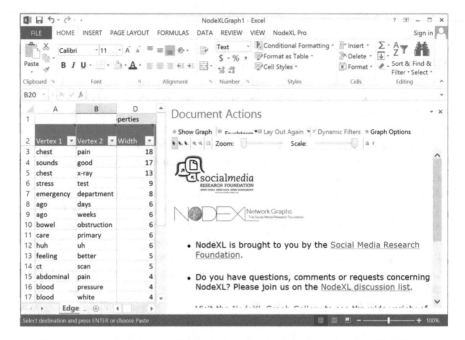

Fig. 5.11 Transfer of the semantic list words pairs and word pair frequency into NodeXL

The next step was to visualize the semantic network using a network visualization tool. The analyst used NodeXL (Smith et al., 2010), which can be downloaded from: https://nodexl.codeplex.com/. An easy way to import the data into NodeXL is to delete the column headers, "source_id," "target_id," and "frequency" from the semantic list, and then copy and paste all the remaining cells in column A of the semantic list into *Vertex 1* under the "Edge" tab in NodeXL. The remaining cells in column B should be pasted under *Vertex 2* (see Fig. 5.11). It is important to make sure that the word pairs match up with one another in the NodeXL spreadsheet the same way they do in the semantic word pair list.

NodeXL's *Width* column is used to display the frequency of each word pair, and it does this visually by adjusting the relative width of the edges linking words in the network map. The analyst copied and pasted the frequency column of the semantic list into the *Width* column in NodeXL. Next, under the "NodeXL" tab at the top of the page, she selected ***AutoFill Columns*** and selected ***Vertex Label*** from the "Vertex" drop down menu and clicked ***AutoFill***. This feature displays the words as labels on the graph. Next, the analyst pressed "Show Graph" in the Document Actions Pane to view the semantic network (Fig. 5.12). An initial network visualization is often uninterpretable because of the many overlapping words and connections. The analyst chose to analyze the underlying structure of the medical consultations network by looking for evidence of subconversations. The procedure

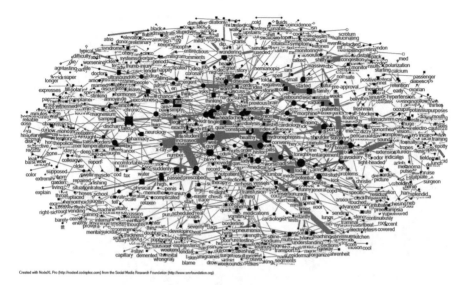

Created with NodeXL Pro (http://nodexl.codeplex.com) from the Social Media Research Foundation (http://www.smrfoundation.org)

Fig. 5.12 The semantic network generated from the medical consultation dataset

used to do this was cluster analysis, which is run by going to the "NodeXL" tab, clicking on ***Groups—Group by Cluster***, and in this case the analyst chose to group the words using the Clauset-Newman-Moore (2004) cluster algorithm. Under the Document Actions Pane she used the layout drop-down menu to select ***Layout Options***, and chose ***Lay out each of the graph's groups in its own box***. Clicking on "Refresh Graph" visualizes the semantic network clusters (Fig. 5.13).

Each of the groups displayed in the visualization of the cluster analysis have been grouped together by the algorithm because the words within each group co-occur with one another more frequently than they do with other words. Each of the groups extracted from the medical consultation dataset represented a conversation topic that arose during the physicians' medical consultations. The analyst examined the individual groups by clicking on the "Groups" tab on the bottom of the NodeXL worksheet, and then clicked on "G1" in the Groups column to highlight the largest group. She exported this group by clicking on ***Export—Selection to New NodeXL Workbook***. This procedure opened up a new NodeXL workbook containing only this group's data. Switching the layout algorithm to Harel-Koren Fast Multiscale (Koren, 2002) and hitting ***Refresh*** made the network structure easier to view. The analyst also clicked on individual words (represented as circular nodes) to adjust the graph image manually so that there were no overlapping or obstructed words. Figure 5.14 shows the subnetwork generated through this process. Figure 5.15 is the second largest subgroup, which was extracted using the same method performed on Subgroup 1.

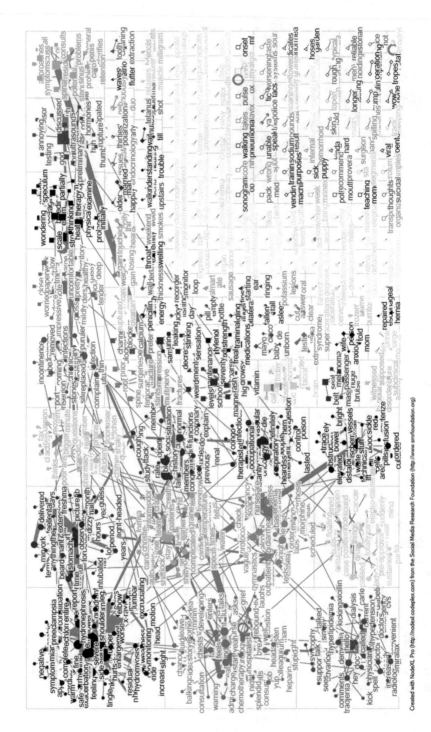

Fig. 5.13 The medical consultation dataset semantic network grouped by cluster

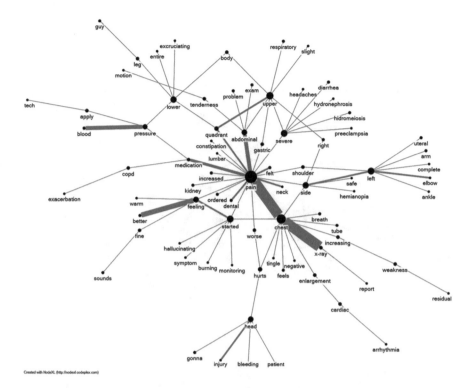

Fig. 5.14 Subgroup 1: Emergency department physician's medical consultations revolving around pain diagnosis and management

5.3.5 Interpretation

When first undertaking interpretation of the results of a semantic network analysis, it is important to remember that during this method, "word associations in texts were analyzed, and those word associations represent[] the meaning inherent to the data" (Doerfel, 1998, p. 23). The resulting graph, such as those in Figs. 5.12 and 5.13, as well as any other metrics or information gained through the analysis, explain something about the relationships between words in the text. *However, the meaning of these relationships can only be gained through interpretation of the results.* For example, finding that the words "sounds" and "good" co-occur frequently within the medical consultations dataset is a meaningless piece of information unless interpretation is done to connect this result back to the data context, the nature of the text archive, and any theoretical frameworks used to collect or interpret the data.

The analyst's interpretation strategy is usually a function of what analyses were performed on the text corpus. This tutorial's example utilized a cluster analysis, and so interpretation of the results will largely focus on interpreting the semantic graphs

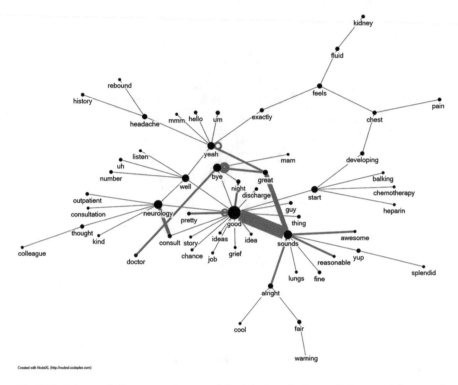

Created with NodeXL (http://nodexl.codeplex.com)

Fig. 5.15 Subgroup 2: Emergency department physician's medical consultations revolving around feedback and affirmation of treatment plans

in terms of what medical consultation conversation topics they indicate. Because so little is known about topics of medical consultations, each conversation cluster should also be evaluated in terms of how these topics manifest within the larger context of the original dataset. As was mentioned earlier, AutoMap linked words when two words existed within the same window frame. While it is likely that words that existed near each other in the text are related in a meaningful way, there is no guarantee that this is the case. Therefore, the prominent, and seemingly meaningful words pairs identified by the network graphs should be searched for within the text corpus to make sure that there are in actuality meaningful relationships between the word pairs.

Some researchers choose to focus on the calculation of graph metrics in order to understand a text corpus, and such metrics should be interpreted in terms of what they explain about the relationships between words or concepts within a dataset. Graph metrics can be calculated at the individual word level (node level metrics) to understand how many connections exist between particular pairs of words. Metrics can also be calculated to understand qualities of the overall graph (graph level metrics). Once again, the simple reporting of a metric like the number of connections between particular words pairs is not enough—the analyst should endeavor to interpret what meaning is indicated by strong or weak connections between word pairs. The researcher may ask: Are there many connections between specific words pairs because they are a common phrase, are they instead two highly-connected concepts, or is there some other reason the words frequently co-occurred? For example, Atteveldt (2008) examined news stories to determine whether words associated with the word "Muslim" changed in news coverage after 9/11. The author found that the word "Muslim" was paired with terrorism-related words in news stories significantly more frequently after 9/11, but that other terror events did not cause an increase in these words' associations. Atteveldt drew on framing theory when interpreting these word associations, finding that "the associative frame between Muslims and terrorism was created not by local events, but rather by 9/11 as a global event" (2008, p. 88).

Just as there are many ways to conduct text mining, there are many approaches to interpreting the results of a text mining study. Overall, the analyst's goal during interpretation should be to: (1) identify patterns generated from the results, (2) confirm that these patterns are true representations of the original text corpus, and (3) interpret these patterns to explain what they represent or mean within the context of the dataset; how they answer a hypothesis or research question; how they can be explained using a theoretical framework; or how the patterns form the grounds for new theory development.

Interpretation of the Medical Consultation Semantic Network Analysis. To briefly review, the results of the text mining and subsequent semantic network analysis revealed the most common communication topics that small groups of physicians in an emergency department discussed as they enacted their expertise to coordinate patient care during medical consultations. These communication patterns were extracted by conducting cluster analysis of word associations within the semantic network. Each cluster contained a group of words that frequently co-occurred with each other and therefore had stronger relationships with one another than they had with other words within the text corpus. The final step of this text mining example is to interpret these patterns.

The largest subgroup (Fig. 5.14), showed the research team that a primary topic of medical consultations for their dataset was the diagnosing and managing of pain. This network graph visually represents all medical consultations in the dataset related to pain. The network graph can be read by starting at the center of the image and tracing the connections outward. In this manner, it is possible to see how consultations regarding chest pain led to the ordering of x-rays and the need for subsequent reports. There are many conversation paths radiating out from the pain node that have to do with describing the exact location of a patient's pain. How pain started and the words patients use to describe the sensation of pain are all parts of this medical consultation topic. From this network, the research team learned about the many ways in which emergency department physicians investigate and treat pain. In terms of the study's goal of understanding physicians' expertise, Subgroup 1 in Fig. 5.14 was interpreted as evidence that the diagnosis and treatment of patients' pain is a primary area of emergency physicians' professional expertise.

This finding was very interesting to the research team because even though they had read through the transcriptions many times, none of the team members had recognized pain as a concept of interest within the dataset. This study illustrates the fact that even though word co-occurrence and frequency are rather simple ways of tallying the presence of words and the relationships between them, this method can help researchers to gain an entirely new perspective of a textual dataset.

The second most dominant pattern found through the semantic network analysis was Subgroup 2 (Fig. 5.15). This subgroup graph displays all conversations that have the phrases "sounds good" or "sounds great" in common, and like Subgroup 1, the graph shows the variations in conversations surrounding these terms. The many other affirmative phrases within this network like "sounds great," "sounds alright," and "yeah" led the research team to interpret this medical conversation topic as evidence of the use of feedback loops by physicians during medical consultation conversations to confirm or affirm treatment plans. The team went back to the text corpus and examined the contexts in which such phrases took place, and this follow-up examination of the text corpus confirmed that these words were very much used by physicians to communicate mutual understanding during medical consultations. This subgroup was interpreted as evidence that feedback is a very important part of enacting expertise during medical consultations. Looking again at the original text archive, the researchers also found that all physician roles, from medical resident to senior physician, utilized these feedback loops, indicating that feedback is an integral component of medical consultation regardless of a physician's level of medical expertise. Although text mining findings primarily originate from analysis of the textual data itself, it is always advisable to collect several layers of information about the context of a textual dataset because this contextual information can help an analyst achieve a more meaningful interpretation of the text mining results.

5.4   Contributions

In this tutorial, text mining aided a research team working to understand physicians' expertise in several meaningful ways. First, the researchers initially hit a roadblock when analyzing their dataset using traditional qualitative thematic coding. The physicians' language contained a lot of jargon, and as outsiders to the medical world, the researchers had a very difficult time finding topical differences that could help them categorize the consultations. This study is also an example of how text mining is useful for small as well as large datasets when barriers exist to traditional analysis methods. The fact that the research team did not notice that pain was a common medical consultation theme when reading the text corpus is further proof of the value of even the simplest text mining procedures.

This study was also the first step towards building theories to explain how physicians enact expertise and how they communicate to manage patient care. Text mining was valuable in helping the researchers take this first step because it allowed them to conduct data-driven discovery in order to identify meaningful conversation topics without having to first develop hypotheses. So little was known about the content of medical consultations that it would have been difficult to form specific hypotheses. Knowing that they were conducting data-driven discovery, the research team carefully defined the scope of their data (medical consultations) and used text mining to explore their data for significant patterns of medical consultation conversations. The research team also conducted follow-up interviews with the physicians they observed during data collection in order to get the physicians' interpretation of the results. The combined quantitative and qualitative results of this study are helping the researchers to build empirically-driven communication and organizational theory. Text mining is also useful for testing theories by looking for patterns within a text corpus to see whether they support existing theory. Additionally, theory can be used as a framework for gathering textual data or for interpreting the results of text mining. Text mining is a very flexible method well suited to making theoretical advancements, but as was discussed earlier, the many choices the researcher makes during data collection, preprocessing, and analysis determine whether or not a text mining analysis ends up being a good fit for a particular research goal like the development of theory.

There are many more text mining procedures and techniques than the few introduced during this tutorial. After discovering that pain and feedback terms were very relevant words in the medical consultation dataset, the research team could conduct further text mining by using these terms as key words, conducting key word analysis in order to extract all words surrounding the words the previous analyses found to be important within the dataset. This approach would tell the research team more about the specific context surrounding these meaningful terms. The research team might be able to learn more about physicians' expertise by having emergency department physicians rate the individual medical consultations according to the level of expertise they represent and then analyze high and low expertise consultations separately in order to evaluate what really excellent consultations have in common and what features are associated with poorly done medical consultations. In a different study it might make sense to take time into consideration, dividing up a text corpus into time segments and analyzing each segment independently in order to understand how a phenomena of interest evolves or develops over time.

There are an infinite number of ways in which to conduct text mining, and this is both a strength of the method and a barrier to its adoption. There is no guarantee that any meaningful results will come from many hours of data formatting, preprocessing, and analysis because the patterns each text mining procedure looks for can be present or absent from a dataset—the analyst cannot know if there is any merit in running a procedure until the work has been invested in running it. The way in which a dataset has been collected also greatly influences the success of text mining. Text mining is often described as an excellent method for analyzing very large text corpuses, but if the text contained within a very large dataset does not have very much in common, text mining is unlikely to identify any patterns, or if it does, the patterns may be more a function of word prevalence within a certain language or context and not due to the existence of important patterns within the data. For example, text mining may find patterns within a text corpus comprised of 10,000 newspaper articles, but if the researcher did not choose newspaper articles that all focus on a specific issue or social phenomena, or if there are off-topic articles mixed in with the corpus, the results of text mining of this data are unlikely to be interpretable in a meaningful way. Even though text mining is a powerful computational tool, it must be combined with good data collection and preprocessing decisions made by a human being who understands exactly what each algorithm and procedure is doing to the data.

Text mining is a very useful tool for both academic research and practical applications in business, education, and individual contexts. It can be used to help analysts learn more about the exponentially-increasing text archives that are generated while we work, from online commenting and debates, through communication with friends and family, and during every online interaction and email we send. The benefits offered by text mining will increase as this method is utilized by people from many disciplines and fields, especially if those who use text mining continue to share the procedures and techniques they find to be useful. Although text mining has existed since the invention of the computer, it is still in its early stages of development and application by people who are not advanced programmers or software engineers. The potential of text mining will increase for everyone as it is adopted for novel applications by new users like readers of this chapter.

References

van Atteveldt, W. (2008). *Semantic network analysis: Techniques for extracting, representing and querying media content*. Charleston, SC: BookSurge Publishing.

Caputo, A., Giacchetta, A., & Langher, V. (2016). AIDS as social construction: Text mining of AIDS-related information in the Italian press. *AIDS Care, 28*, 1171–1176.

Carley, K. (2001). AutoMap (version 3.0.10.41) [Computer software]. Pittsburg, PA: CASOS, Carnegie Mellon University. Retrieved from http://www.casos.cs.cmu.edu/projects/automap/index.php

Carley, K. M., Columbus, D., Bigrigg, M., Diesner, J., & Kunkel, F. (2010). AutoMap User's Guide 2010 (CMU-ISR-10). Carnegie Mellon University. Retrieved from http://www.casos.cs.cmu.edu/publications/papers/CMU-ISR-10-121.pdf

Choudhary, A. K., Oluikpe, P. I., Harding, J. A., & Carrillo, P. M. (2009). The needs and benefits of text mining applications on post-project reviews. *Computers in Industry, 60*(9), 728–740.

Clauset, A., Newman, M. E. J., & Moore, C. (2004). Finding community structure in very large networks. *Physical Review E, 70*.

Cohen, A. M., & Hersh, W. R. (2005). A survey of current work in biomedical text mining. *Briefings in Bioinformatics, 6*(1), 57–71.

Collier, N., Doan, S., Kawazoe, A., Goodwin, R. M., Conway, M., Tateno, Y., … Shigematsu, M. (2008). BioCaster: Detecting public health rumors with a Web-based text mining system. *Bioinformatics, 24*(24), 2940–2941.

Dasgupta, S., & Sengupta, K. (2016). Analyzing consumer reviews with text mining approach: A case study on Samsung Galaxy S3. *Paradigm, 20*(1), 56–68.

Doerfel, M. L. (1998). What constitutes semantic network analysis? A comparison of research and methodologies. *Connections, 21*(2), 16–26.

Fan, W., Wallace, L., Rich, S., & Zhang, Z. (2006). Tapping the power of text mining. *Communications of the ACM, 49*(9), 76–82.

Kessler, C. S., Afshar, Y., Sardar, G., Yudkowsky, R., Ankel, F., & Schwartz, A. (2011). A prospective, randomized, controlled study demonstrating a novel, effective model of transfer of care between physicians: The 5 Cs of consultation. *Academic Emergency Medicine*, *19*, 968–974.

Koren, D. H. Y. (2002). A fast multi-scale method for drawing large graphs. *Journal of Graph Algorithms and Applications*, *6*(3), 179–202.

Lammers, J. C., Lambert, N. J., Abendschein, B., Reynolds-Tylus, T., & Varava, K. (2016). Expertise in context: Interaction in the doctors' room of an emergency department. In P. M. Leonardi, & J. W. Treem (Eds.), *Expertise in Organizations* (pp. 145–167). Oxford: Oxford University Press.

Lexalytics. (2015). Dealing with context in text mining [White paper]. Retrieved August 29, 2016, from Lexalytics: https://www.lexalytics.com/content/whitepapers/Lexalytics-WP-Context.pdf

Luhn, H. P. (1958). The automatic creation of literature abstracts. *IBM Journal of Research and Development*, *2*(2), 159–165.

Maughan, B. C., Lei, L., & Cydulka, R. K. (2011). ED handoffs: Observed practices and communication errors. *American Journal of Emergency Medicine*, *29*, 502–511.

Miner, G. (2012). *Practical text mining and statistical analysis for non-structured text data applications*. New York: Academic Press.

Smith, M., Ceni A., Milic-Frayling, N., Shneiderman, B., Mendes Rodrigues, E., Leskovec, J., & Dunne, C. (2010). NodeXL: A free and open network overview, discovery and exploration add-in for Excel 2007/2010/2013/2016. Retrieved from http://nodexl.codeplex.com/

Chapter 6
Sequential Synchronization Analysis

Toshio Murase, Marshall Scott Poole, Raquel Asencio, and Joseph McDonald

6.1 Introduction

Sequences have long been a central interest in group research.[1] Sequences capture how group processes unfold over time, and characterization of sequences as a whole and their properties offers valuable insights into group decision-making, conflict management, group cohesion, teamwork, and many other group phenomena.

Sequences have been studied on a variety of levels in group research. Some of the best known sequences are the stages of the group life cycle. While Tuckman's (1965) iconic "Forming, Norming, Storming, and Performing" stage sequence is the best known of these, several dozen models of the group life course have been described (Hare, 1976, 2010; LaCoursiere, 1980). Sequential models of specific group activities such as problem solving (Bales & Strodtbeck, 1951), decision making (Fisher, 1970; Poole & Roth, 1989), conflict (Pondy, 1967), and teamwork

[1] Preparation of this chapter was supported by National Science Foundation grant #BCS 0941268 and Army Research Institute grant W5J9CQ-12-C-0017. The contents of this chapter represent the opinions of the authors and not of these organizations.

T. Murase (✉)
Department of Psychology, Roosevelt University, Chicago, IL, USA
e-mail: toshio.murase@gmail.com

M.S. Poole
University of Illinois, Urbana, IL, USA
e-mail: mspoole@illinois.edu

R. Asencio
Krannert School of Management, Purdue University, Lafayette, IN, USA
e-mail: rasencio@purdue.edu

J. McDonald
Human Interfaces, Inc., Austin, TX, USA
e-mail: joe@humaninterfaces.net

© Springer International Publishing AG 2017
A. Pilny, M.S. Poole (eds.), *Group Processes*, Computational Social Sciences,
DOI 10.1007/978-3-319-48941-4_6

(Ishak & Ballard, 2012; Marks, Mathieu, & Zaccaro, 2001) have also been advanced. Conceptually, these activity sequences can be thought of as embedded within longer group life cycles. Still other scholars have focused on short cycles of group activity that might be repeated multiple times within episodes of group work, such as Tschan's (1995) orientation-action-evaluation cycles, which are posited to be tied to quality of group work.

In studying sequences, researchers can focus on the entire sequence, as did Tuckman (1965), Bales and Strodtbeck (1951), and Poole and Roth (1989). Relevant research questions include: Do all groups follow the proposed sequence?; What factors determine whether a given sequence occurs?; Is following the sequence related to outcomes such as effectiveness and group cohesion? A second option it so focus on subsequences that make up the entire sequence, as Tschan (1995) and Murase et al. (2015) did. In this case relevant questions include: What types of subsequences occur and what is their frequency?; How do they chain together to generate longer sequences and what types of longer sequences occur?; How are they related to outcomes such as group effectiveness or group cohesion? Finally, researchers may identify characteristics of sequences or subsequences, such their frequency, complexity (Poole & Roth, 1989), or conformity to an ideal sequence (Poole & Roth, 1989) or subsequence (Tschan, 1995). Relevant research questions are: How do various sequences compare in terms of the properties?; What factors govern variability in the characteristics?; How do the characteristics relate to outcomes such as group effectiveness or group cohesiveness?

The approaches described in the previous paragraph focus on the sequence as a property of the group as a whole. Another approach is to decompose the sequential data from the group level to the individual level. In this case the sequence of behaviors of each member is analyzed. Just as with group level sequences, individual sequences can be characterized in terms of their overall structure, subsequences, and characteristics, and the same questions posed for the group as a whole can be posed for the sequences of individual members. But decomposition also enables researchers to explore the processes that lead to the emergence of a group or its properties from the interactions among members.

One of the oldest questions in group research is "What makes a group more than just a collection of individuals?" There has been a long debate over whether a group has an entitivity beyond the behaviors of its individual members (Davis, 1969; Hewes, 1996; Kozlowski, Chao, Grand, Braun, & Kuljanin, 2013; Kozlowski & Klein, 2000). Kozlowski and Klein (2000) argue that higher level group properties emerge through two processes, *composition* of individual attitudes, knowledge, and/ or behaviors into aggregates and *compilation*, which depends on nonlinear combination of individual attitudes, knowledge and/or behaviors. McGrath and Kelly (1986) and Ancona and Chong (1996) consider temporal elements of coordination among individual member sequences. *Entrainment* is defined as cases in which the pace, rhythm, and cycles of individual behaviors come into alignment with one another. In this case, the group's activity takes on a character of a holistic unit greater than the individual members. McGrath and Kelly argued that entrainment depends on an external factor such as the group's task or a leader or events in the environment that the group must respond to. However, it also seems possible that entrainment might also be driven by members' desire to coordinate and engage one another in internal group interaction. The study of synchronization and entrainment of member behavior enables us to investigate the degree to which the group transcends individual member activities.

This chapter will provide an overview of several methods for sequence analysis that address these questions, including whole sequence methods, short cycle methods, and sequential synchronization analysis. Methods for whole sequence and short cycle analysis have been discussed at length elsewhere, so they will be described in general terms; sequential synchronization analysis has not been previously introduced, so the remainder of the chapter will be devoted to an explanation of how it works and can be conducted.

6.2 Sequence Analysis

6.2.1 Sequence Data

Group sequence data can come from a number of sources. It can be directly recorded by observers (e.g., Bales, 1950), or it can be coded from audio or video recordings (e.g., Fisher, 1970; Poole, 1981). Researchers like Axelrod (1976, 2015) used archives of diplomatic notes and negotiations to reconstruct sequences of argument. Data can also be gathered using computerized group or team simulations of, for example, military tasks, emergency patients, or negotiations (e.g., Schiflett, Elliott, Salas, & Coovert, 2004), which capture automatically the choices and actions of each member down to hundredth of a second units. Another data resource for group research is data captured from the internet (e.g., email, social media, text messages) and mobile devices (e.g., geolocation, sociometric badges).

Figure 6.1 presents a general illustration of the type of sequence data that results from the operations described in the previous two paragraphs. The top row shows the basic data units. These units are then coded into meaningful categories (in this case A, B, C, and D), which are the elements of the sequence. As the previous discussion shows, in some cases the coding system defines the units as part of the coding process (e.g., Interaction Process Analysis), while in other cases (e.g., a military simulation) the units are "hard-coded" into the data recording apparatus, while in still others (e.g., server data from a massive multiplayer online game) the units must be retrieved from a more complex data store. Each unit may also be associated with a timestamp, shown in the bottom row of the figure; this timestamp orders the elements and may also be used to determine durations. The timestamp in this figure is based on a "Newtonian" conception of time, in which time can be divided into equal units and proceeds linearly into the future. The top row of the figure portrays a different conception of time, "event time," in which the occurrence of events marks the units, regardless of how long they were or the intervals between them. In addition to time stamps, this data also indicates the source of or major actor in each unit. Note that a member may engage in several consecutive acts.

Units (U*)/Members (M*):

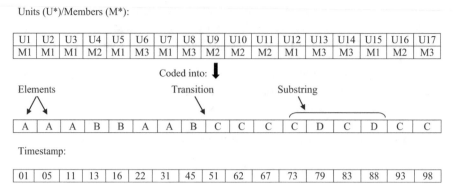

Fig. 6.1 Sequences and sequence data

Some properties of sequence data are shown in the second row of Fig. 6.1, transitions from one element to the next. Substrings (or subsequences) are meaningful short-term patterns of acts; they may be defined structurally by repeated sequences of elements or theoretically by specification of meaningful sequences of elements (e.g., plan-act-evaluate). Identification of meaningful units or subsequences sometimes proceeds through a series of hierarchical steps. As the third row of the figure indicates, each series of similar units can be re-coded into a single occurrence or phase of this unit. A phase is a coherent period of group activity of the same type. In this case, the phasic sequence is ABABCDCDC. This can be reduced to a still higher-order pattern, as shown in row four, in which repeating AB substrings are reinterpreted as E phases and CD substrings as F phases. Poole and Roth (1989) used this approach to simplify phase sequences in group decision-making using a procedure formally described in Holmes and Poole (1991).

6.2.2   Analyzing Sequences

Many group process studies analyze sequences by "collapsing" them into profiles of the total number of each type of act in the sequence. These profiles are useful because they show general differences between sequences. A sequence with a lot of conflict events is clearly different from one with very few.

Information is lost, however, by synoptic measures of processes such as profiles. Where in a decision process a conflict occurs tells us a lot about the process. A conflict early on may serve to raise issues for the group to discuss and resolve; a conflict at the end may create an impasse that stymies the group. Considering the sequence of activities tells us the "story" of the group process in a way that simple totals cannot.

Rudimentary sequence analysis has often been applied to coded data in the social sciences. Human pattern recognition is powerful and adaptive, making it possible to extract rich information about human interaction and behavior from video sessions (e.g., DeChurch & Marks, 2006; Kozlowski, Chao, Chang, & Fernandez, 2015; Stachowski, Kaplan, & Waller, 2009). Bales and Strodtbeck (1951), for instance, divided their discussions into thirds and constructed graphs of amounts of orientation, evaluation, and control behavior over time to compare sequences of group problem solving sessions. However, if researchers have sequences made up of many units or a large set of sequences, manually identifying critical patterns is a difficult and daunting task. Methods developed in the biological sciences to identify DNA sequences from millions and billions of data points (Koonin & Galperin, 2003) and in computer science, where strings of thousands of digits or lines of code must be compared (Sankoff & Kruskal, 1983) can be brought to bear in this case. To overcome this challenge, these disciplines developed approaches to data mining and large scale analytics designed to find unique patterns of information and to evaluate similarities in structure and function between sequences (Needleman & Wunsch, 1970).

Sequence analysis is particularly aligned with process models that posit that groups develop through a series of distinct stages (Tuckman, 1965) and engage in patterns of phases to make decisions and accomplish their tasks (Bales & Strodtbeck, 1951; Gersick, 1988; Poole & Holmes, 1995; Poole & Roth, 1989; Sambamurthy & Poole, 1992). For example, Marks et al. (2001, see also Ishak & Ballard, 2012) proposed temporally-based team process in which team members engage in two types of phases alternatively to achieve objectives: transition phases—where members engage in planning and strategizing—and action phases—where they engage in activities directly contributing to team performance.

Sequence analysis is also appropriate for models of act-to-act sequences. The assumption commonly shared among these models is that events and behaviors trigger each other to create unique contexts in which one leads to another, which then facilitates the occurrence of more events and behaviors later on (Lehmann-Willenbrock, Meyers, Kauffeld, Neininger, & Henschel, 2011). Tschan's (1995) plan-act-evaluate behavioral cycle model of effective team activity is a good example of this approach.

6.2.2.1 Whole Sequence Analysis

Poole and his colleagues investigated the phasic sequences groups followed to make decisions (Poole & Holmes, 1995; Poole & Roth, 1989; Sambamurthy & Poole, 1992). Instead of measuring members' perceptions of their decision-making process, Poole and Roth (1989) content-coded 47 decision processes by taking the following steps: (a) identifying major activity (e.g., problem-focused, execution-focused, and solution-focused activities) within each 30-second time segment of a process to create a sequence of the activities; (b) grouping into phases the activities of the same category if they occurred consecutively and also grouping into phases activities from the different categories if they happened in a row. They used the technique of flexible phase mapping (Holmes & Poole, 1991) to identify various sequences and methods including optimal matching to compare and classify sequences into types. The sequences of activity phases produced by this method provided the fine-detailed picture of when the specific activity phases occurred and in what order. For example, some groups always went through a fixed process of different phases while others moved through different stages and cycled back to the previous stages. The richness of the sequence data helped Poole and Roth uncover that groups did not follow unitary group process but that their processes were much more complex and diverse.

One useful technique in whole sequence analysis is optimal matching (OM), which is designed to compare similarities of pairs of sequences (Abbott & Tsay, 2000; Aisenbrey & Fasang, 2010; Hollister, 2009; Wu, 2000). OM evaluates how similar pairs of sequences are. It assesses the degree of difference (distance) between pairs of sequences using substitution-insertion-deletion transformation operations (INDEL). Suppose one wants to compare two sequences: ABC and ADE. OM calculates the distance between them by using the INDEL operations. First, OM replaces B at the second position of ABC with D; inserts E between D and C of sequence ADC, which turns the sequence into ADEC; and then deletes C at the last position of the sequence. The number of the operations required to convert the first sequence into the second one is 3, which is the distance score between these two sequences. Weights are generally attached to various INDELs based on similarity of elements. For example of A and B both pertain to problem statements and to a solution statement, substituting A for B would make less difference than substituting C for A, So the B-A substitution would be given lower weight (cost) than the C-A substitution. Based on this logic optimal matching algorithms assign weighted differences to each pair of sequences in a set. The number of ways to calculate distance scores between a pair of long sequences increases drastically. Therefore, OM seeks the most optimal ways to calculate distance scores among sequences (Abbott & Tsay, 2000).

The resulting set of distance scores can then be analyzed using multidimensional scaling or clustering techniques to derive sets of sequences with similar structures. For example, Sambamurthy and Poole (1992) derived three different sets of sequences from a sample of 45 conflict management discussions: one in which conflict was suppressed, one in which there were open disagreements that were not resolved, and a third in which there was open discussion and cooperative management of the conflict. The third set had more positive relationships to outcomes than the other two. It is also possible to take a reference sequence—for example, an ideal type sequence—and use optimal matching to determine how similar one or more sequences are to the reference sequence.

There has been much debate over the proper use and benefits and costs of using OM. Readers can refer to Aisenbrey and Fasang (2010) and Herndon and Lewis (2015) for further discussion of these issues.

6.2.2.2 Subsequence Analysis

While Poole and colleagues studied entire sequences, Lehmann-Willenbrock et al. (2011) examined whether mood emerges through short-cycles of behavioral patterns in which complaining behavior leads to supporting behavior which leads to complaining behavior. They coded discussions in which 57 company teams discussed solutions to problems in their work activities. Each statement provided by an employee in the conversation was assigned to one of 44 behavioral categories, resulting in a sequence of behaviors for the team. Lehmann-Willenbrock et al. examined how often one behavioral type was followed by another by calculating probability ratings among all possible pairs of behaviors in the 44 × 44 table. Using the probability ratings, they found that team members often engaged in specific cycles of complaining behaviors (e.g., complaining, complaining, and complaining; complaining, supporting, and complaining), and that the cycles of complaining behaviors resulted in unaroused and unpleasant group mood while the cycles of positive behaviors produced pleasant group mood. Methods such as relational event modeling can be used to test hypotheses about short cycle sequences as well (see Chap. 4, this volume).

Murase et al. (2015) took a different approach to obtain sequences of actions from six-person teams participating in a military simulation game. The server recorded in milliseconds various acts which team members performed, producing sequence data consisting of thousands of thousands of acts over time. Murase et al. developed 37 behavioral categories important for the game, each of which contained short sequences of acts that occurred in specific orders. They then wrote scripts to count the number of times subsequences of acts in the log that matched any of the 37 behavioral categories occurred (they employed 30 s windows for sampling purposes). Their sequence data showed which member in the team engaged in what type of behavior in which time segment. This data was subsequently used in an analysis of social entrainment among team members that will be described in the next section.

Poole, Lambert, Murase, Asencio, and McDonald (2017) and Cornwell (2015) summarize these and other sequence analysis techniques, along with theoretical and data related issues. The bibliographies of these two works list a number of references to more detailed descriptions of specific sequence methods. The remainder of this chapter focuses on the method of sequential synchronization analysis, which facilitates identification of emergent processes such as teamwork through the coordination of the behavioral streams of individual members.

6.3 Sequential Synchronization Analysis

6.3.1 Individual Sequences into Group Processes

To conduct sequential synchronization analysis the researcher first decomposes the group sequence into a sequence for each member and then analyzes relationships among individual data sequences to determine team level dynamics.

Two theoretical forms have been advanced to explain how group dynamics emerge at the team level: compositional and compilational models (Chan, 1998; Kozlowski & Klein, 2000; Roberts, Hulin, & Rousseau, 1978). Compositional models argue that a phenomenon at the individual level resembles the same form of the phenomenon at the team level while compilational models argue that the forms of a phenomenon at the individual and team level are different.

Compositional models are based on the logic that each member's behavior can serve as an estimate of the group or team's behavior, because the phenomenon of interest manifests in the same way at the individual and group levels. Averaging the individual estimates thus yields a more reliable measure of the group or team's behavior. For example, in the case of group decision-making, information sharing is such that any information given by a single member can be used by the entire group. So it makes sense to take each members' information sharing (or, in the case of self-report measures, perceptions of group information sharing level) and combine or average them to get an overall measure for the group.

In contrast, compilational models operate under a logic of individual variability that assumes that it is the pattern or variation among members that gives the group process its character (Murase, Doty, Wax, DeChurch, & Contractor, 2012). So, if one member of a team is quarrelsome and difficult, this can disrupt the team's activity no matter what other members do. Or members may specialize, as in a transactive memory system, where one member specializes in remembering past mistakes and serves as devil's advocate, while another specializes in coming up with novel ideas to address the problems raised by the first. Only if the group has individual members who enact these and other key roles, will it make an effective decision. So it is the pattern of members rather than any sort of sum total that characterizes the emergent group, and to capture this emergence, the various types of patterns or at least variance among members must be characterized. Measures for compilation include the standard deviation, minimum and maximum score of the team members, or gini coefficients on various measures such as personality traits, self-efficacy, or member roles (Barrick, Stewart, Neubert, & Mount, 1998; Campion, Medsker, & Higgs, 1993; Stewart, Fulmer, & Barrick, 2005). All of these measures are based on individual characteristics of members or synoptic, summary measures of group interaction, rather than the group process itself. One influential theory that offers a process-oriented, nonsynoptic account of group emergence from individual activities is the theory of social entrainment (Ancona & Chong, 1992; McGrath & Kelly, 1986).

6.3.2 Entrainment

A great deal of evidence suggests that human behavior—including group and team behavior—is patterned by rhythms and temporal cycles. McGrath and Kelly (1986) summarize evidence that human interaction is characterized by "complex temporal patternings of multiple sets of responses by multiple social actors. These patterns have been expressed by such terms as 'mutuality,' 'reciprocity,' 'complementarity,' 'dominance,' 'similarity,' 'simultaneity,' and 'alternation'" (p. 7). Cappella (1991) makes a case that at the dyadic level these rhythms and patterns in interaction are biologically determined. Poole and Roth (1989) noted that about 40 % of decision-making groups engaged in repetitive cycles of problem-solution interaction. Tschan (1995) showed that short repetitive cycles of problem-solving were characteristic of effective teams.

McGrath (1990) argues that activities in social systems operate in rhythmic and cyclic forms. Multiple activities, initially operating in different rhythms, eventually get locked into the same rhythmic pattern by influencing one another's pace or adjusting their activity rhythms to the rhythms of dominant members or external events. For example, project deadlines, unexpected requests from a client, and a competing company's market entry function as dominant rhythms to which members on teams must adjust their work paces (Ancona & Chong, 1992). Once the activities have settled into a fixed rhythmic pattern, it becomes persistent even when the dominant activity ceases, unless another disrupting event or new dominance pacer emerges to which the activities must start entraining (Harrison, Mohammed, McGrath, Florey, & Vanderstoep, 2003). These studies have demonstrated that synchronization of activities among members is a mechanism underlying the emergence of group-level phenomena.

Most previous research has relied on experimental manipulations and/or measurement of members' perceptions to capture synchronization. However, it is also possible to identify synchronization from behavioral sequences.

For example, to accomplish a specific objective in a military team exercise, members may increase the level of a relevant behavior (e.g., attacking an enemy unit). Once the objective has been accomplished, the level of the behavior begins to decrease and then eventually cease for a while. This cycle repeats as triggering events (new enemy combatants) occur. In this case, members engage in oscillating activity patterns with one cycle representing a basic behavioral unit, defined as a peak-to-peak period (Cazelles & Stone, 2003). The overlap degree of peak-to-peak periods between pairs of activity cycles essentially determines synchronization degree and type.

If the peaks of multiple members' oscillating patterns occur at the same time points, or the pace in which the peaks occur is the same (regardless of whether or not the peaks occur at the same time points), those members are said to be entrained to one another (Ancona & Chong, 1992). Ancona and Chong define the former as *synchronic entrainment* and the latter as *tempo entrainment*. If the peaks of pairs or sets of the oscillating patterns at completely at the alternating points, they are defined as *harmonic entrainment*. Figure 6.2a, b demonstrate two types of entrainment where pace is defined as a period from one peak (maximum) at t time to another peak at $t+1$ time of a cycle (Cazelles & Stone, 2003). Various statistical measures of the properties of pairs or sets of patterns—discussed below—can be used to determine whether various types of entrainments hold in a group.

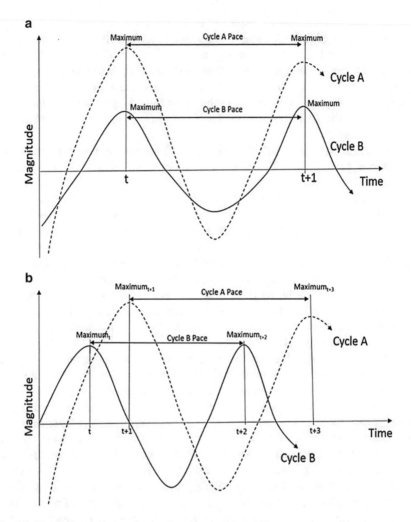

Fig. 6.2 Types of entrainment: (**a**) synchronic entrainment, (**b**) temporal entrainment

6.4 A Step-by-Step Guide to Sequential Synchronization Analysis

This section is organized to provide step-by-step directions for identifying sequences and then calculating phase-lock scores from a hypothetical time-series data, which are used to capture the degree of synchronization of team behavior. The approach to identifying sequences was used in Murase et al.'s study (2015) which counted frequencies of sequences using the R package TraMineR (Ritschard, Bürgin, & Studer, 2013) and calculated phase-lock scores using the R package synchrony (Gouhier & Guichard, 2014).

We provide a hypothetical study in which four members participate in a military simulation game in which two four-member teams must navigate a course through enemy positions. In order to do perform effectively, their units collect and exchange information important to their mission and also coordinate attacks on enemy units. There are eight events in this scenario: (A) collecting information, (B) member's unit health decrease, (C) attack, (D) enemy health decrease, (E) communication, (F) enemy death, (G) exchanging information, (H) moving with other member, (I) moving alone, and (J) moving close to the enemy. These elements are documented at one-second intervals in the order in which they occurred during the hypothetical mission. The data set is available for download for those who are interested in analyzing it at http://hdl.handle.net/2142/91573. The R code for conducting the analysis is referenced below in the example.

The dataset is made up of two teams of four members each. Each row represents a series of events performed by a single member. In this data, the events B, D, and F1 appear across all the members when any of these events occurs to at least one member because they are events that happen to or have impacts on all members of both teams. For example, Member 1 on Team 1 starts engaging the enemy at the 19th position, and the enemy's health decreases at the 21st position. Although this event belongs to Member 1, it is documented across all the members, because the enemy's health decrement is beneficial for any member who encounters this enemy. This irregular, "messy" data structure is typical of sequence data sets, particularly those derived from digital traces. This underscores the value of attending to temporal patterns in data rather than individual acts: focusing on event D alone for Members 2, 3, and 4 might lead us to conclude incorrectly that these members engaged the enemy; but focusing on the sequence CD (attack → enemy health decrease) for Member 1 uncovers the meaning of the event, showing that the result for all was a product of the Member 1's action.

The methods discussed in this section can be applied to simple units like those just defined or to more complex units such as subsequences. In our discussion we will use subsequences as our basic unit of analysis, on the premise discussed in the previous paragraph, that using subsequences or cycles as basic units gives us a more nuanced and accurate description of member behavior.

6.4.1   Step 1: Theoretically Define the Units of Interest

The first and most important step is to develop a set of theoretically sound units of analysis. When using single acts, the coding system often specifies them. In the case of subsequences construct definition occurs through considering meaningful combinations of acts. Not all subsequences are necessarily meaningful and even when all are, only a few might be of interest given the theory being tested. These serve as basic units of analysis. One challenge lies in the process of putting events in specific orders to create sequence bases because theories in social sciences typically do not specify sets of events and in what exact order those events should unfold. It is the researcher's responsibility to carefully evaluate what events and behaviors need to be included and in what order they should be placed so that the short sequences can capture the concepts of interest.

For example, in team research, explicit and implicit coordination have been found to influence team performance (Rico, Sánchez-Manzanares, Gil, & Gibson, 2008). Explicit coordination is defined as the process in which members communicate to define responsibilities, make plans and deadlines, and exchange information in order to orchestrate their efforts and activities to achieve common objectives. On the other hand, implicit coordination emphasizes members' ability to predict each other's activities in the process of orchestrating their efforts (Rico et al., 2008). As can be seen, these definitions do not precisely specify what exact behavioral events should be included and in what order. The researcher must choose the behaviors that fit these definitions.

The subsequences of implicit and explicit coordination can include combinations of several different types of behavioral events. For example, using the categories defined above, one subsequence for explicit coordination starts from communication with member A to moving with member A to being close to enemy. On the other hand, a subsequence of implicit coordination starts from moving along to moving with member A to being close to enemy because the definition of implicit coordination emphasizes one's ability to predict other members' behavior (Rico et al., 2008). This definition suggests that communication should not be the essential part of short sequences which capture implicit coordination.

Additionally, the researcher must determine how long the subsequences should be. An appropriate length should be long enough so that below that length a sequence of events should not be complete, but above it a sequence can be broken down into smaller subsequences. For example, it is difficult to determine what type of construct can be captured by a subsequence of two behaviors which starts from moving alone to communication, because depending on what behavioral events come before or after this sequence, the meaning of the sequence changes. If the events, moving with other team and being close to enemy, come after this sequence, the new subsequence with the four events could mean explicit coordination. One member tells another member nearby that he is moving toward the enemy unit, and asks the member to come to his location. Then these two meet and move together toward the enemy unit. If these two behaviors do not come after the original sequence, it can be too short to determine whether it captures explicit coordination or something else.

On the other hand, if a subsequence is too long, it could consist of two or more subsequences, each of which alone could provide sufficient information to capture a theoretical construct. For example, if a sequence is assumed to consist of six actions of moving alone, communicating, moving with another member, being close to enemy, attacking, and enemy health decrease, this sequence can be broken into the first subsequence of four behavioral elements—moving alone, communicating, moving another member, and being close to enemy—and a second subsequence of attacking and enemy health decrease. The first subsequence is explicit coordination, and the second subsequence defines a new construct: engaging enemy. Therefore, the researcher must consider not only the *"what events"* question (what events need to be included) but also the *"how many events"* question (how many events are necessary to make one complete sequence).

Furthermore, the researcher can create multiple subsequences all of which can belong to the same construct. There is no reason to expect that there should be only one subsequence per construct. For example, psychological scales are comprised of multiple items because having multiple questions is considered necessary to capture different aspects of the same construct (Nunnally & Bernstein, 1994). This perspective can be applied to the sequence-based method. If one subsequence may not be enough to capture the entire construct space, multiple subsequences are necessary to obtain adequate coverage of the construct.

This first step is essential for ensuring legitimacy for this type of method. It is common in computer science to simply mine sequences and use the obtained set. However, if we want to relate our sequence analysis to theory, this "dustbowl empiricist" approach would not be sufficient. For the eight act categories we had above, there would be 56 possible pairs for each individual team member and many more if we consider three and four act sequences. This is simply too many to sort through. Generating the subsequences of interest based upon both theory and empirical findings from the literature provides a solid framework through which the researcher can appropriately interpret the meanings of subsequences uncovered by data mining. Without theoretical guidance, the researcher will be easily overwhelmed by the enormous number of short sequences identified through data mining alone.

Out of hundreds of possible sequences, Murase et al. (2015) defined seven different subsequence types comprised of 37 actual subsequences to represent four key teamwork constructs: implicit coordination, explicit coordination, taskwork, and information gathering. Two subsequence types indicated implicit coordination, two explicit coordination, two taskwork, and one indicated information gathering. In this case they used teamwork theory to guide a multilevel classification scheme that started with 37 meaningful sequences, which were then grouped into seven basic types, which were then mapped onto the four key teamwork constructs.

6.4.2   Step 2: Extract Subsequences from Data

The next step is to extract subsequences of events from the longer sequence of each participant. The R package TraMineR (Gabadinho, Ritschard, Mueller, & Studer, 2011) can be used to conduct a number of different types of sequence analyses. TraMineR contains numerous R functions with which researchers can create and manipulate data for sequence analysis, mine data to find unique sequences, and visualize results. Researchers who are more familiar with Stata can conduct similar types of sequence analysis using Stata packages such as SAID (Halpin, 2014) and others (e.g., Brzinsky-Fay, Kohler, & Luniak, 2006). The rest of the analytical demonstration will be conducted using TraMineR.

In this case we want to extract subsequences from the data. While we know theoretically which subsequences we are looking for, it is useful to mine the full set of subsequences for additional information. In some cases, additional unanticipated subsequences that correspond to our theoretical constructs may be identified. In other cases one or more subsequences might suggest additional constructs compatible with our theoretical orientation.

To extract subsequences, we use the subsequence function (which is called seqefsub in the Synchrony package) to mine event sequences in the form of shifts from one type of behavior to another type. One consideration is subsequence length. The length of a sequence could be anywhere from 2 units (i.e., A → B) to the entire length of data collected in one's study. A second consideration is how to deal with repeats of the same unit multiple times in a row. When data are documented in every second as they are in a game, the same event can be recorded for a member many times in row; for instance, if the player is moving continuously, then movement will be recorded each second so long as the continuous movement occurs. As a result, the data can contain a long string of the same events with a different element at the end (i.e., AAAAAAAB), and the repeats are an artifact of the recording. The subsequence function identifies no shift (A) and one shift from A to B (A→B) at the end, and ignores the intervening multiple occurrences of the element.

When we employ a subsequence identification technique like seqefsub that only identifies shifts from one type of act to another (and ignores successive repeats of the same unit), we recommend that the researcher consider whether to break data into multiple shorter segments to limit the time period over which subsequences can extend. If the original sequence runs over hours, months, or days, techniques like seqefsub might identify subsequences which extend over longer stretches of time than humans can realistically act over or attend to. If one's sequence data spans 60 min, for example, mining the entire sequence makes no sense because the subsequence function will pull out many sequences which are not meaningful. For example, the function could identify a shift between two behaviors—communication with member A during the first 30 s of the session and moving with member A 25 m into the game. Such a shift does not make sense given the nature of teamwork interaction patterns, in which members typically respond relatively immediately to one another. To avoid this issue, the researcher should consider breaking the time into multiple time segments within which shifts between units are considered meaningful. The appropriate length of time segments will vary according to the phenomenon. A reasonable latency period for teamwork is relatively short, while in the case of organizational innovation adoption sequences could extend over days, weeks, or months and still be meaningful.

The second decision point is to determine how many shifts are allowed to be part of short sequences. The subsequence function could completely exhaust the entire list of short sequences, and it could take significant computing resources to complete the identification process if the empirical sequence is very long. For more efficient subsequence identification, the researcher should determine the appropriate number of shits which are maximally allowed in short sequences. If too many shifts are allowed, they would not be interpretable or can be broken down into shorter sequences. In our case, we limit the length of sequences to be no more than 3 shifts (i.e., A→B, B→C, C→D), which is in line with the decisions on this matter made by other researchers (Lehmann-Willenbrock et al., 2011; Murase et al., 2015; Poole & Roth, 1989).

The last decision point is to consider how far apart the behaviors within the same shift or the shifts within the same sequence are allowed to be. Suppose that there is a sequence of As and Bs at 10 positions (AAABAABBAA) and that the researcher is interested in identifying the short sequence (A→B)−(B→A). First, the researcher considers whether the events of the same shift should occur at the positions right next to each other or at the positions somewhat apart from each other. For example, it is important to consider whether A_1 and B_4 (the subscripts indicate the event positions in the sequence) are allowed to define a shift or whether only adjacent acts like A_3 and B_4, and A_6 and B_7 should be identified as shifts. The same concern must be exercised when the researcher considers which shifts should be included in the same subsequence. Depending on how far apart the behaviors within the same shift and shifts within the same short sequence are allowed to be located, the subsequence function produces different frequencies even for the same short sequence.

To operationalize various choices related to relationships among units in subsequences, there are several different counting operations one can use: one occurrence per object (COBJ), one occurrence per span-window (CWIN), distinct occurrences with possibility of event-timestamp overlap (CDIST_O), and distinct occurrences with no event-timestamp overlap allowed (CDIST) (Joshi, Karypis, & Kumar, 1999).

COBJ counts a specified sequence only once throughout the entire data even if the sequence appears more than once. This is an appropriate rule to use when once a subsequence occurs its full effect is felt. CWIN uses a moving window within which it evaluates the occurrence of the short sequence. First, the researcher must determine how many units a moving window covers every time it moves. For example, if the moving window is set to cover three units, every time it moves, it assesses whether the sequence occurs in those three units. After the moving window goes through the entire data set, the CWIN function provides the total number of occurrences of the short sequence. This rule is appropriate if every occurrence of the subsequence counts. Finally, CDIST_O identifies all possible short sequences within the window whose length is specified by the researcher. The CDIST_O function differs from CDIST in that CDIST counts only one occurrence of the short sequence in a window, whereas CDIST_O counts all occurrences within the window, even those that overlap. More detailed descriptions and comparisons of the counting operations can be found in Joshi et al. (1999).

6.4.3 Step 3: Revisit Theoretically Defined Subsequences in Light of Sequence Mining Results

The subsequence functions CWIN and CDIST_O will identify all possible combinations of subsequences and count their frequencies. In step 1 the researcher makes the decisions that define the types of subsequences that will be identified.

Table 6.1 Counts of subsequences

Set of events		Base sequence	Short sequence 1	Short sequence 2
A, B	Sequence	(A→B)	(A→B) – (B→A)	(A→B) – (A)
	Frequency	7	6	7

Note: The arrow sign indicates a shift from one behavior to another; a hyphen connects two shifts.

No theory allows the researcher to make perfect determinations about all meaningful subsequences that indicate theoretical constructs. Additional promising subsequences may have been identified in the sequence mining process. The next task, then, is to use these results to refine the subsequence indicators that are supposed to capture the target constructs. Only those subsequences which indicate the target constructs or suggest new constructs that fit within the theoretical framework should be retained and all the rest should be discarded. Although this process seems straightforward, it is not.

Table 6.1 presents a scenario with the set of events which any short sequences identified must contain. For example, two other short sequences contain the set of AB events and provide their frequency information. Note that two letters connected by the arrow consist of a shift while the hyphens connect two shifts to create a longer chain. Suppose you have identified A and B as critical events, and the subsequence function has identified two other subsequences (A→B) – (B→A) and (A→B) – (A→B). The issue faced in this scenario is that the two latter chains contain the A→B shift as part of their sequence so you wonder how this information can be combined. Because of the same A→B shift in the both short sequences, their frequency counts are not independent of each other but are redundant. As you can see, the base sequence (A→B) occurred seven times. This means that any short sequences containing the base sequence can occur more than seven times. Thus, unless, the specific short sequence (A→B) – (B→A) is the target short sequence whose occurrence is 6, the researcher should record 7 for this scenario while discarding the other frequency numbers.

As the length of the original sequence data increases, the number of subsequences one can make exponentially increases and becomes impossible to count manually. Utilizing the data mining approach provides the researcher with the new ability to capture information that the researcher cannot think of without the data mining technique.

6.4.4   Step 4: Aggregate Frequency Counts of Subsequences for Data Segments

In step 2 we argued that any long sequence could be broken into shorter segments that reflect realistic latencies in thought and action and also ease computational demands. Once an appropriate set of subsequence indicators have been identified, the next step is to count them in each segment to yield a sequence of counts for each individual member. Carrying through our example of the categories discussed in Step 1 this would yield values of the number of subsequences devoted to explicit coordination, implicit coordination, taskwork, and information gathering for each segment. The result is four time series, one for each activity, for each member.

6.4.5   Step 5: Compute Synchronization Scores

Entrainment can be assessed by calculating the degree and type of synchronization across the individual member time series. The output of values of the algorithm provides a means for calculating the degree to which members remain phase-locked or socially entrained throughout the game. Suppose two members have coordination cycles with the same pace. If they coordinate with each other at the same time throughout the game, the cycle value differences are zero. However, even if their paces are the same, members can engage in coordination at different time points. For example, one member coordinates in every 5 min at the 5th, 10th, and 15th minute, but the other member engages in coordination at the 3rd, 8th, and 13th minute. In this case, the cycle value differences yield a series of non-zero constants. Finally, if members engage in coordination at random time points and change the pace of these cycles, the cycle differences yield a series of random numbers. It is important to note that this third scenario represents members who are not entrained to one another.

Because the phase-lock algorithm produces random numbers for non-entrained members, the phase-lock calculation can determine the degree to which members are entrained by the distribution of the previously calculated cycle differences, with uniformly-distributed values representing low phase-lock (i.e., low entrainment) (Cazelles & Stone, 2003). For every pair of members, cycle difference scores for every time point are calculated to create a distribution. If two members' coordination cycles are in perfect sync, the cycle difference scores are zeros while two members that constantly and randomly change their pace would create a uniform distribution of the difference scores. Therefore, if the distribution of cycle differences has a clear peak, two members are said to be "phase-locked", and if the distribution spreads out and approaches uniformity, phase-lock decreases. We use kurtosis values to represent the degree of "peakedness" of cycle-difference distributions.

Besides the phase-lock technique which is the main synchrony analysis in this chapter, other synchrony analysis techniques which are also available in the synchrony package deserve attention. Community-wide synchrony (Loreau & de Mazancourt, 2008) evaluates the degree to which members' time-series data fluctuate in unison. Kendall's coefficient concordance is a non-parametric statistic which evaluates agreement among members' time-series data (Gouhier, Guichard, & Gonzalez, 2010). Although these statistics can be used to evaluate entrainment, the phase-lock technique is the most appropriate because it capture similarity between peak-to-peak paces of multiple cycles, which we used to define entrainment. When using other techniques, we recommend that researchers carefully consider the definition of entrainment and then select the most appropriate technique.

6.5 Example

In this section we analyze the sample dataset mentioned earlier. Table 6.2 summarizes ten basic activity elements team members engaged in the game and describes of which of three coordination sequences the actions should be part. These three sequences are also specified in the Searchcode file at http://hdl.handle.net/2142/91572. This file currently allows readers to specify up to six elements that sequences should and should not contain. Elements that the sequences must contain need to be specified in the "action" columns, and TRUEs must be specified in the "yesno" column. If there are some elements that should not be part of sequences, they must be specified in the action columns, and FALSEs must be specified at the appropriate positions in the yesno columns. For example, the first row in the Searchcode file contains A and G and two TRUEs, meaning that mined sequences must contain A and G. If sequences should not contain, for example, G, the TRUE at the second position should be changed to FALSE. If A should not be contained, the first TRUE should be changed to FALSE.

Two R code scripts for sequence and synchrony analysis are available for download at http://hdl.handle.net/2142/91573. The scripts help readers understand how we prepared data for sequential synchrony analysis and conducted the analyses. It is difficult to provide the full description in this chapter for what we did line-by-line given limited space, but we attempt to highlight the main lines important for the analysis and provide explanations. The further explanations for all the script lines are provided directly in the scripts.

We broke the data into 10 20-second time segments as we recommend in the Step 2 section, and identified all sequences within each segment to create time-series data per member. The code to create the time segments is shown in Table 6.3.

Table 6.2 Coding categories used in the example

Element	Action	Coordination sequence: engaging enemy	Coordination sequence: exchanging information	Coordination sequence: planning
A	Pick up information			✓
B	Health decrease			
C	Attack	✓		
D	Enemy health decrease			
E	Communication		✓	
F1	Enemy death		✓	
G	Exchange information			✓
H	Moving with other team	✓		
I	Moving alone			
J	Move close to enemy	✓		

Note: Check mark indicates of which sequence the element/action is part

Table 6.3 Dividing data into segments

2	seg<−20
12	**for**(time_i **in** c(1:((ncol(teamdat)-3)/seg))){
13	print(paste("MTS",mts_i," Mem",member_i," Seg",time_i,sep=""))
14	lst<−time_i*seg
15	fst<−lst+1−seg
16	subdat<−teammemdat[1,fst:lst]
17	variability<−length(unique(apply(subdat,2,as.character)))

Line 2: Object seg indicates the 20-second time window used to divide the data into ten different time segments
Line12: The for-in function specified the number of time the lines that follow should repeat
Lines 14–15: The beginning and ending of each time segment are calculated
Line16: Object subdat is the data segment extracted from a member's entire data set (teammemdat)

Next we identified sequences within each segment. First, the CDIST counting operation was used to identify sequences that contained up to three shifts. Once identified, sequences were evaluated for whether they captured team coordination, and their frequency counts were documented if they contained one of the sets of behaviors in the following order: H, J, and C; F1 and E; A and G. These three sets of behaviors indicate different ways in which members engage in team coordination. Sequences containing H, J, and C indicate that members move together to engage enemy. Sequences containing F and E indicate that members plan for the next move after they complete a task (which is removing the enemy threat). Finally, sequences containing A and G indicate that members exchange information as they locate it. Although we could generate more combinations of behaviors, we use only these three sequences in this demonstration. If sequences contained any other behaviors which were not specified in this section, their frequency counts were not documented. Table 6.4 shows the commands given to TraMineR for this operation.

The next step was to examine whether sequences members engaged in within the same time segments were considered as redundant or unique. For example, Member 2 on MTS 1 engaged in three sequences containing H, J, and C in the seventh time segment: (H) − (H→J) − (J→C); (H) − (J→C); and (H→J) − (J→C). If the frequencies of all the three sequences were included, the total count for this segment would be 3. However, if the chain of actions in this segment is evaluated, it is obvious that these three are actually duplicates. The chain is HHHHHHHHHJJCCCCBDCDBDC. This member engaged in this type of coordination activity only once in this time segment as indicated in that the member engaged in one series of move activities and one series of attack activities. Therefore, we took only one sequence out of these three and documented its frequency count. Furthermore, we took this approach through the entire data. This is a complex operation that is explained in the code available for download.

Additionally, when members engaged in different types of coordination within the same time segments, we took the sum of their frequencies. For example, Member 2 on MTS 1 engaged in two different types of sequences in the eighth segment: (H→J) − (J→C) and (F1→E), and each sequence occurred only once. The reason

Table 6.4 Code for CDIST

18	**if**(variability>1){
19	datsize<−ncol(subdat)
20	eve.seq<−seqdef(subdat)
21	eve.seqe<−seqecreate(eve.seq)
22	fsubseq<−seqefsub(eve.seqe,minSupport=1,maxK=3,
23	constraint = seqeconstraint (maxGap = datsize,
24	windowSize = datsize,countMethod='CDIST'))
25	evecount<−matrix(seqeapplysub(fsubseq,method="count"),
26	ncol = 1, nrow = ncol(seqeapplysub(fsubseq,method="count")))

Line18: The if function checks whether the segment contains different action elements
Line20: The seqdef function converts a character string into a vector of events
Line21: The seqecreate function prepares data for sequence analysis from the event sequence object created in Line 20
Line22: The seqefsub function mines the data to produce sequences with maxK=3 meaning the function creates sequences of up to three shifts. CDIST is selected as counting operation for this analysis
Line25: The seqeapplysub function produces data containing frequency counts for the sequences found in Line 22. This line prepares the frequency count data in the matrix format

for this approach being adequate is that on average the members engaged in coordination sequences only once in each segment. Thus, summing frequencies of different types of sequences did not distort team coordination information. However, this approach could produce distorted information if frequencies for one type of sequences were exceedingly larger than those for the other types of sequences, but all the types of sequences were considered equally important. For example, in some data teams typically engage in implicit coordination about 100 times with standard deviation (SD) of 20 while engaging in explicit coordination 10 times with SD of 2. Additionally, we assume the researchers consider these two types of coordination equally important. However, if the frequency counts of these two types are summed across, the aggregate score that is supposed to represent the coordination construct is over-represented by implicit coordination, which is not aligned with how this construct is conceptualized. In that case, researchers could convert frequencies into z-scores first and take sum of them. Fortunately, in the current data, this was not a concern.

Table 6.5 summarizes the frequency counts of sequences that met the aforementioned criteria. Member 1 and 4 on Team 1 did not engage in activities as much as the other members while all the members on Team 2 were active throughout all the time segments.

In the last step, calculation was conducted on the extent to which members' activities over time were phase-locked. Using R package Synchrony (Gouhier & Guichard, 2014), phase-lock scores were calculated for every pair of members within each team, and then kurtosis scores were derived to evaluate the degree of peakness (Table 6.6).

Table 6.5 Frequency counts of coordination sequences over Time

Team	Member	Time 1	Time 2	Time 3	Time 4	Time 5	Time 6	Time 7	Time 8	Time 9	Time 10
1	1	0	1	0	0	0	1	0	1	0	1
1	2	1	1	1	1	0	0	1	2	1	1
1	3	0	1	0	2	1	0	1	2	0	0
1	4	0	0	1	0	0	0	0	0	0	1
2	1	1	1	0	1	0	2	1	2	1	1
2	2	2	1	1	1	1	2	2	1	2	0
2	3	0	1	0	1	0	2	2	1	0	1
2	4	0	1	1	1	0	2	1	1	1	0

Table 6.6 Synchrony commands

23	**for**(mem_ii **in** unique(mtsdat[,"member_i"])
24	[−1*which(unique(mtsdat[,"member_i"])==members)])
25	{
26	run<−run+1
27	t1<−as.numeric(mtsdat[mtsdat[,"member_i"]==mem_i,−1:−2])
28	t2<−as.numeric(mtsdat[mtsdat[,"member_i"]==mem_ii,−1:−2])
29	div_i<−length(unique(t1))
30	div_ii<−length(unique(t2))
31	**if**((div_i>=3)|(div_ii>=3)){
32	sync.maxs<−phase.sync(t1,t2,mins=TRUE)
33	k<−0
34	s<−NA
35	ave<−NA
36	sds<−NA
37	k<−kurtosis(sync.maxs$deltaphase$mod_phase_diff_2pi,na.rm.=TRUE)
38	s<−skewness(sync.maxs$deltaphase$mod_phase_diff_2pi,na.rm.=TRUE)
39	ave<−mean(sync.maxs$deltaphase$mod_phase_diff_2pi,na.rm.=TRUE)
40	sds<−sd(sync.maxs$deltaphase$mod_phase_diff_2pi,na.rm.=TRUE)

Lines 27–28: This line selects a pair of members (mem_i and mem_ii) from the team data (mtsdat)
Lines 29–30: This line calculates the numbers of actions the members performed, and then calculates synchrony scores for the pairs whose numbers of actions are equal to or more than 3
Line32: The phase.sync function calculates the synchrony scores between two members' time-series data and automatically creates a distribution of cycle differences of the synchrony scores
Line37: The kurtosis function calculates the degree of peakedness from the distribution of the cycle differences
Line38: The skewness function calculates how skewed the distribution is

Table 6.7 Kurtosis scores used to evaluate synchronization

Team	Pair	Kurtosis
1	1 and 2	2.73
1	1 and 3	1.84
1	1 and 4	NA
1	2 and 3	1.70
1	2 and 4	1.63
1	3 and 4	NA
2	1 and 2	1.50
2	1 and 3	2.23
2	1 and 4	NA
2	2 and 3	2.72
2	2 and 4	1.50
2	3 and 4	1.50

Table 6.7 summarizes kurtosis scores across all the pairs of members among the two teams, with higher scores indicating the more peaked the cycle difference distribution becomes (Cazelles & Stone, 2003). Values closer or larger than 3 indicate that the distribution has a higher peak than the normal distribution, which indicates that two members are entrained to each other. From this table, the 1–2 pair on Team 1, and the pairs of 1–3 and 2–3 on Team 2 have values closer to 3, indicating that their distributions have a higher peak than the normal distribution (DeCarlo, 1997). Interestingly, the kurtosis value between Member 1 and 2 was higher than that between Member 2 and 3. Although Member 2 and 3 were more active than the other members, Member 1 and 2 had more synchronization on their activities than did the other pair. Another notable point is that the phase-lock calculation produced NAs for the pairs involving Member 4. Member 4 was inactive as evidenced in that this member engaged in coordination only twice. Calculating phase-lock values requires enough fluctuation in data so it may not be useful if one's data contain many members being inactive throughout.

6.6 Discussion

In this chapter, we have provided a step-by-step guide to perform sequence synchrony analysis to investigate the degree to which team members are socially entrained. Specifically, there are two objectives of the chapter. The first objective is not simply to explain how to use specific R functions from the R packages "synchrony" and "TraMineR", but how to evaluate the theoretical relevance of behavioral elements that should be part of subsequences. The hybrid method of data mining and theory-based thinking provides a solid foundation on which subsequences mined from data acquire substantive meaning and relevance to one's study. The second objective is to provide a further guidance on how to obtain unique team property "social entrainment" from subsequence data rather than simply calculating average scores across members. By combining these two methods, sequential synchrony analysis enables researchers to capture compilational forms of emergence.

Group properties emerge in compilational and compositional forms as individuals become cohesive functioning teams (Chan, 1998; Kozlowski & Klein, 2000). Although researchers have argued importance for compilational forms, they have mainly relied on compositional forms or taking average scores to capture team properties. This practice suggests that the current state of science on group and team process is limited because the most preferred analytical approaches are designed to capture only compositional forms. We argue that a reason for the lack of utilizing compilational forms is that there is no theoretical as well as analytical guide to capture them. To spur the use of compilational forms, we have attempted to develop a solution to both of the problems.

Past studies have effectively demonstrated sequence analysis as a powerful technique in preserving contextual meanings of team processes. Sequence analysis can capture compilational forms of emergence especially when researchers directly conduct sequence analysis on data at the team level to obtain patterns of interactions in the team (Lehmann-Willenbrock et al., 2011; Poole & Roth, 1989; Tschan, 1995). However, this technique alone is not sufficient to capture compilational forms when it is conducted on individual-level time-series data because it simply converts the meaning of data from the raw information to subsequences. As a result, the converted data still require aggregation to be elevated to the team level. This is the situation we have illustrated in the example, where researchers must have a specific theoretical and analytical guide to obtain compilational forms.

Social entrainment (McGrath & Kelly, 1986) is a theoretical framework that serves a guide when researchers wonder what team property emerges at the team level in a compilational form. Social entrainment takes on a compilational form when it emerges because each member's behavioral rhythm does not accurately depict how synchronized members' behaviors are. One useful way to observe this phenomenon is to conduct synchronization analysis on members' time-series data.

Like all sequential process analysis, sequence synchronization analysis is a "work in progress." Currently, there are no definitive, canonical techniques for process analysis as there for analysis of experimental designs. While these are emerging, at this point sequence analysis requires improvisation and ingenuity. We encourage readers to build on what we have described as they pursue their own projects.

References

Abbott, A., & Tsay, A. (2000). Sequence analysis and optimal matching methods in sociology review and prospect. *Sociological Methods & Research, 29*, 3–33.

Aisenbrey, S., & Fasang, A. E. (2010). New life for old ideas: The "second wave" of sequence analysis bringing the "course" back into the life course. *Sociological Methods & Research, 38*, 420–462.

Ancona, D. G., & Chong, C. (1996). Entrainment: Pace, cycle, and rhythm in organizational behavior. In L. L. Cummings, & B. M. Staw (Eds.), *Research in Organizational Behavior* (vol. 18, pp. 251–284). Greenwich, CT: JAI Press.

Ancona, D. G., & Chong, C. L. (1992, August). Timing is everything: Entrainment and performance in organization theory. In *Academy of Management Proceedings* (Vol. 1992, No. 1, pp. 166–169). Academy of Management.

Axelrod, R. (1976). *Structure of decision*. Princeton, NJ: Princeton University Press.

Axelrod, R. (Ed.) (2015). *Structure of decision: The cognitive maps of political elites*. Princeton, NJ: Princeton University Press.

Bales, R. F. (1950). *Interaction process analysis; a method for the study of small groups*. Oxford, England: Addison-Wesley Press.

Bales, R. F., & Strodtbeck, F. L. (1951). Phases in group problem-solving. *The Journal of Abnormal and Social Psychology, 46*, 485–495.

Barrick, M. R., Stewart, G. L., Neubert, M. J., & Mount, M. K. (1998). Relating member ability and personality to work-team processes and team effectiveness. *Journal of Applied Psychology, 83*, 377–391.

Brzinsky-Fay, C., Kohler, U., & Luniak, M. (2006). Sequence analysis with Stata. *Stata Journal, 6*, 435–460.

Campion, M. A., Medsker, G. J., & Higgs, A. C. (1993). Relations between work group characteristics and effectiveness: Implications for designing effective work groups. *Personnel Psychology, 46*, 823–847.

Cappella, J. N. (1991). The biological origins of automated patterns of human interaction. *Communication Theory, 1*(1), 4–35.

Cazelles, B., & Stone, L. (2003). Detection of imperfect population synchrony in an uncertain world. *Journal of Animal Ecology, 72*, 953–968.

Chan, D. (1998). Functional relations among constructs in the same content domain at different levels of analysis: A typology of composition models. *Journal of Applied Psychology, 83*, 234–246.

Cornwell, B. (2015). *Social sequence analysis: Methods and applications* (vol. 37). Cambridge: Cambridge University Press.

Davis, J. H. (1969). *Group performance*. Reading, MA: Addison-Wesley.

DeCarlo, L. T. (1997). On the meaning and use of kurtosis. *Psychological Methods, 2*, 292–307.

DeChurch, L. A., & Marks, M. A. (2006). Leadership in multiteam systems. *Journal of Applied Psychology, 91*, 311–329.

Fisher, B. A. (1970). Decision emergence: Phases in group decision-making. *Communications Monographs, 37*, 53–66.

Gabadinho, A., Ritschard, G., Mueller, N. S., & Studer, M. (2011). Analyzing and visualizing state sequences in R with TraMineR. *Journal of Statistical Software, 40*(4), 1–37.

Gersick, C. J. (1988). Time and transition in work teams: Toward a new model of group development. *Academy of Management Journal, 31*(1), 9–41.

Gouhier, T. C., Guichard, F., & Gonzalez, A. (2010). Synchrony and stability of food webs in metacommunities. *The American Naturalist, 175*, 16–34.

Gouhier, T. C., & Guichard, F. (2014). Synchrony: Quantifying variability in space and time. *Methods in Ecology and Evolution, 5*, 524–533.

Halpin, B. (2014). Three narratives of sequence analysis. In *Advances in sequence analysis: Theory, method, applications* (pp. 75–103). Springer International Publishing.

Hare, A. P. (1976). *Handbook of small group research* (2nd ed.,). New York: Free Press.

Hare, A. P. (2010). Theories of group development and categories for interaction analysis. *Small Group Research, 41*, 106–140.

Harrison, D. A., Mohammed, S., McGrath, J. E., Florey, A. T., & Vanderstoep, S. W. (2003). Time matters in team performance: Effects of member familiarity, entrainment, and task discontinuity on speed and quality. *Personnel Psychology, 56*, 633–669.

Herndon, B., & Lewis, K. (2015). Applying sequence methods to the study of team temporal dynamics. *Organizational Psychology Review, 5*, 318–332.

Hewes, D. E. (1996). Small group communication may not influence decision-making: An amplification of socio-egocentric theory. In R. Y. Hirokawa, & M. S. Poole (Eds.), *Communication and group decision-making* (2nd ed., pp. 179–212). Thousand Oaks, CA: Sage.

Hollister, M. (2009). Is optimal matching suboptimal? *Sociological Methods & Research, 38*, 235–264.

Holmes, M. E., & Poole, M. S. (1991). Longitudinal analysis. In B. M. Montgomery, & S. Duck (Eds.), *Studying interpersonal interaction* (pp. 286–302). New York: Guilford Press.

Ishak, A. W., & Ballard, D. I. (2012). Time to re-group: A typology and nested phase model for action teams. *Small Group Research, 43*, 3–29.

Joshi, M. V., Karypis, G., & Kumar, V. (1999, May). A universal formulation of sequential patterns. In *Proceedings of the KDD'2001 workshop on Temporal Data Mining*.

Koonin, E. V., & Galperin, M. Y. (2003). Principles and methods of sequence analysis. In *Sequence—Evolution—Function, Springer US*, pp. 111–192.

Kozlowski, S. W. J., & Klein, K. J. (2000). A multilevel approach to theory and research in organizations: Contextual, temporal, and emergent processes. In K. J. Klein, & S. W. J. Kozlowski (Eds.), *Multilevel theory, research, and methods in organizations: Foundations, extensions, and new directions* (pp. 3–90). San Francisco, CA: Jossey–Bass.

Kozlowski, S. W., Chao, G. T., Chang, C. H., & Fernandez, R. (2015). Team dynamics: Using "big data" to advance the science of team effectiveness. In S. Tonidandel, E. King, & J. Cortina (Eds.), *Big data at work: The data science revolution and organizational psychology*. New York, NY: Routledge.

Kozlowski, S. W., Chao, G. T., Grand, J. A., Braun, M. T., & Kuljanin, G. (2013). Advancing multilevel research design capturing the dynamics of emergence. *Organizational Research Methods, 16*, 581–615.

Lacoursiere, R. B. (1980). *The life cycle of groups*. New York: Human Sciences Press.

Lehmann-Willenbrock, N., Meyers, R. A., Kauffeld, S., Neininger, A., & Henschel, A. (2011). Verbal interaction sequences and group mood exploring the role of team planning communication. *Small Group Research, 42*, 639–668.

Loreau, M., & de Mazancourt, C. (2008). Species synchrony and its drivers: Neutral and nonneutral community dynamics in fluctuating environments. *The American Naturalist, 172*, 48–66S.

Marks, M. A., Mathieu, J. E., & Zaccaro, S. J. (2001). A temporally based framework and taxonomy of team processes. *Academy of Management Review, 26*, 356–376.

McGrath, J. E. (1990). Time matters in groups. In J. Galegher, R. E. Kraut, & C. Egido (Eds.), *Intellectual teamwork: Social and technological foundations of cooperative work* (pp. 23–61). New York: Lawrence Erlbaum Associates, Inc..

McGrath, J. E., & Kelly, J. R. (1986). *Time and human interaction: Toward a social psychology of time*. New York: Guilford Press.

Murase, T., Asencio, R., McDonald, J., Poole M.S., DeChurch, L. A., & Contractor, N. (2015). The effect of synchronization of group processes on multiteam system effectiveness. Presented at the annual meeting of National Communication Association, Las Vegas, NV.

Murase, T., Doty, D., Wax, A. M. Y., DeChurch, L. A., & Contractor, N. S. (2012). Teams are changing: Time to "think networks". *Industrial and Organizational Psychology, 5*, 41–44.

Needleman, S. B., & Wunsch, C. D. (1970). A general method applicable to the search for similarities in the amino acid sequence of two proteins. *Journal of Molecular Biology, 48*, 443–453.

Nunnally, J., & Bernstein, I. H. (1994). *Psychometric therapy*. New York, NY: McGraw-Hill.

Pondy, L. R. (1967). Organizational conflict: Concepts and models. *Administrative Science Quarterly, 12*, 296–320.

Poole, M. S. (1981). Decision development in small groups I: A comparison of two models. *Communication Monographs, 48*(1), 1–24.

Poole, M. S., & Holmes, M. E. (1995). Decision development in computer-assisted group decision making. *Human Communication Research, 22*, 90–127.

Poole, M. S., & Roth, J. (1989). Decision development in small groups IV a typology of group decision paths. *Human Communication Research, 15,* 323–356.

Poole, M. S., Lambert, N. J., Murase, T., Asencio, R., & McDonald, J. (2017). Sequential analysis of processes. In H. Tsoukas, & A. Langley (Eds.), *The SAGE handbook of organizational process studies.* Thousand Oaks, CA: Sage.

Rico, R., Sánchez-Manzanares, M., Gil, F., & Gibson, C. (2008). Team implicit coordination processes: A team knowledge–based approach. *Academy of Management Review, 33,* 163–184.

Ritschard, G., Bürgin, R., & Studer, M. (2013). Exploratory mining of life event histories. In J. J. McArdle, & G. Ritschard (Eds.), *Contemporary issues in exploratory data mining in the behavioral sciences* (pp. 221–253). New York: Routledge.

Roberts, K. H., Hulin, C. L., & Rousseau, D. M. (1978). *Developing an interdisciplinary science of organizations.* San Francisco: Jossey-Bass.

Sambamurthy, V., & Poole, M. S. (1992). The effects of variations in capabilities of GDSS designs on management of cognitive conflict in groups. *Information Systems Research, 3,* 224–251.

Sankoff, D., & Kruskal, J. B. (Eds.) (1983). *Time warps, string edits, and macromolecules: The theory and practice of sequence comparison.* Reading, MA: Addison-Wesley.

Schiflett, S. G., Elliott, L. R., Salas, E., & Coovert, M. D. (Eds.) (2004). *Scaled worlds: Development, validation, and applications.* Hants: Ashgate.

Stachowski, A. A., Kaplan, S. A., & Waller, M. J. (2009). The benefits of flexible team interaction during crises. *Journal of Applied Psychology, 94,* 1536–1543.

Stewart, G. L., Fulmer, I. S., & Barrick, M. R. (2005). An exploration of member roles as a multi-level linking mechanism for individual traits and team outcomes. *Personnel Psychology, 58,* 343–365.

Tschan, F. (1995). Communication enhances small group performance if it conforms to task requirements: The concept of ideal communication cycles. *Basic and Applied Social Psychology, 17,* 371–393.

Tuckman, B. W. (1965). Developmental sequence in small groups. *Psychological Bulletin, 63,* 384–399.

Wu, L. (2000). Some comments on sequence analysis and optimal matching methods in sociology: Review and prospect. *Sociological Methods & Research, 29,* 41–64.

Chapter 7
Group Analysis Using Machine Learning Techniques

Ankit Sharma and Jaideep Srivastava

7.1   Machine Learning Techniques and Tools

Our aim in the following text is to provide a hands-on experience for group researchers to use machine learning and data-mining methods. Our main focus is to analyze and understand variables that may affect the group's performance. Keeping that in mind we shall illustrate the use of two machine learning and data-mining methods in a variety of combinations for group performance analysis. We employ an existing implementation of these methods in data-mining GUI based software named Weka (Hall et al., 2009). We shall also illustrate the process of moving from individual level variables to group level metrics in the Data Description Section. In the next subsections we describe the methods (Decision Trees and Feature Selection methods) and introduce the Weka tool.

7.1.1   Decision Trees

In machine learning, decision trees were first introduced by Quinlan (1986) in form of the ID3 algorithm. Later, Quinlan (1993) proposed the C4.5 algorithm to improve upon the limitation of ID3 algorithm. The major improvements upon ID3 are (1) C4.5 can handle both discrete as well as continuous data, (2) it can also handle missing data, and (3) C4.5 also does tree pruning. In the following chapter we shall be using the C4.5 algorithm for building the decision trees because of these reasons.

A. Sharma (✉) • J. Srivastava
University of Minnesota, Minneapolis, MN, USA
e-mail: sharm170@umn.edu; srivasta@cs.umn.edu

© Springer International Publishing AG 2017   145
A. Pilny, M.S. Poole (eds.), *Group Processes*, Computational Social Sciences,
DOI 10.1007/978-3-319-48941-4_7

Table 7.1 Training samples of 14 days with two features and dependent variable as team played or not that day

#	Outlook	Humidity	Play
1	Sunny	High	No
2	Sunny	High	No
3	Rainy	High	Yes
4	Rainy	High	Yes
5	Rainy	Normal	Yes
6	Rainy	Normal	Yes
7	Sunny	Normal	Yes
8	Sunny	High	No
9	Sunny	Normal	Yes
10	Rainy	Normal	Yes
11	Sunny	Normal	Yes
12	Sunny	High	Yes
13	Rainy	Normal	Yes
14	Rainy	High	No

Decision trees are supervised learning methods that make use of already classified training data to build predictive models. The aim of a decision tree classifier is to divide the training samples into partitions that are homogeneous with respect to the dependent variable (which in our analysis would be the group's performance). The algorithm outputs a model in the form of a tree where the bottom or end nodes (*leaves*) are the final predictions (or the classification class) and all the other nodes (*non-leaves*) represent some independent variables. During the construction of a tree, that independent variable is chosen as the node which splits its set of samples in the most homogeneous fashion i.e. each split is homogeneous with respect to the dependent variable. For this, the C4.5 algorithm employs a normalized information gain (Quinlan, 1993) as the criterion for variable selection and the variable with the highest normalized information gain (i.e., best predictor) is chosen as the node.

As an example we have 14 samples where each sample has a day's humidity and outlook and depending upon these variables if a group plays a cricket game or not, given in the Table 7.1. Using the C4.5 implementation in Weka software we achieve the decision tree shown in the Fig. 7.1b. If we look at the tree, the root is chosen as "humidity" by the algorithm and not the "outlook" variable. To understand this, if we try to split the days if the team will play or not, on the basis of the values of "outlook" and "humidity" variables individually, we get splits as shown in Fig. 7.1a. As we can see that if "humidity" variable is "normal" then we get a split of seven instance days on which the group always plays. In this sense, this split generated by "humidity" variable is pure i.e. all the instances are "yes" only. This purity is what we have been referring to as homogeneous split. Given that "humidity" is able to generate a more homogeneous split we say it is a more informative variable and thus, choose it over the "outlook" variable. Right now for illustration purposes we diagrammatically illustrated the splits and just by eye balling we can understand which split is homo-

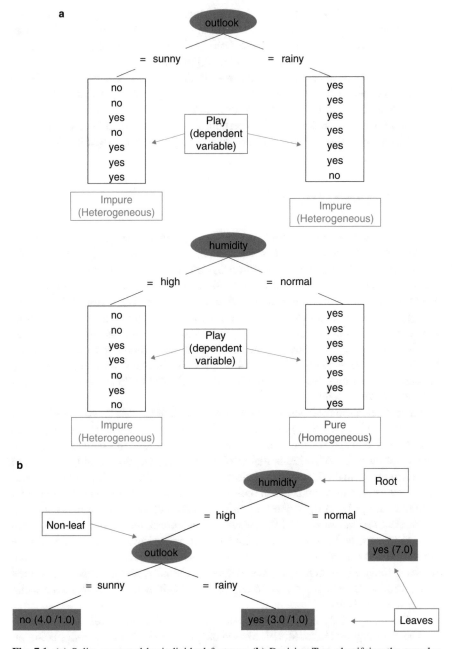

Fig. 7.1 (**a**) Splits generated by individual features. (**b**) Decision Tree classifying the samples from Table 7.1

geneous or more informative or not. However, this is impractical in practice and C4.5 employs an information theoretic measure of normalized information gain (Quinlan, 1993) as the criterion for variable selection. For further details of this measure we encourage readers to visit the Quinlan's text (Quinlan, 1993).

The biggest advantage of decision trees is that a single tree has the ability to describe the whole feature space. This ease of interpretability makes them quite popular among practitioners and therefore, we propose them for social scientists as a tool to understand the feature space pertaining to groups. We make use of an open source implementation of this algorithm available in the Weka software we use.

7.1.2   Feature Selection

Given the training samples, the aim of *feature selection* is to select a compact subset of independent variables that can predict the dependent variable without much loss of information. In other words, the purpose is the trim the dataset into a manageable one by focusing on independent variables that have high predictive power. Feature selection mines the most informative features and gets rid of the redundant or strongly correlated features. This process helps achieve a compact smaller set of features (i.e., parsimony) and therefore, improves model interpretability as well as training time and generalization by less over fitting (modal selection) (Guyon, Saffari, Dror, & Cawley, 2010). For a general overview of feature selection in machine learning we refer to (Guyon & Elisseeff, 2003) and the survey (Chandrashekar & Sahin, 2014).

Feature selection methods are mainly categorized into three types: (1) Filter, (2) Wrapper and (3) Embedded (Guyon et al., 2010). A subset of features can be judged as informative or not irrespective of how well they are able to predict the target or dependent variable. Algorithms that perform feature selection in this manner are called Filtering methods but as the selection is independent of the prediction accuracy, they usually may not perform optimally. Wrapper methods evaluate the model accuracy using a learning method for different subset of features and return the best performing feature subset. But the evaluation and search are done separately, making wrapper methods often computationally expensive. Embedded methods, on the contrary try to merge the subset search and evaluation phase, by incorporating the search within the machine learning model itself. Therefore, the information obtained while training the model are used to eliminate or retain features, all this done while model training itself.

In this paper we describe the application of a popular embedded method called SVM-RFE (Support Vector Machine based on recursive feature elimination) (Guyon Weston, Barnhill, & Vapnik, 2002). This algorithm reclusively learns SVM based model and eliminates independent variables or features with low weights. For further details of the algorithm we refer the reader to the original paper in (Guyon et al., 2002). We make use of the open-source implementation of SVM-RFE in Weka, which is called "SVMAttributeEval".

7.1.3 Introducing WEKA: GUI Based Machine Learning Tool

We conduct analysis using the tool called Waikato Environment for Knowledge Analysis (Weka), written in Java and developed at University of Waikato, New Zealand. This is a free software available for Windows, Linux as well as Macintosh environments at (Hall et al., 2009). The tool's website has link to numerous tutorials and they also have video based courses at YouTube. The best part of tool is the easy Graphical User Interface (GUI) which makes it very popular among data-mining and machine learning practitioners.

7.2 Dataset and Metrics

7.2.1 Dataset Collection and Description

The dataset was collected using a game based test-bed: SABRE - Situation Authorable Behavior Research Environment, developed by BBN Technologies, using the Bioware's Neverwinter Nights game and its provided toolset (Leung, Diller, & Ferguson, 2004). In this research we employ a NATO dataset collected using the game-based test-bed (SABRE) (Fig. 7.2). During the experiment 56 teams, of four members

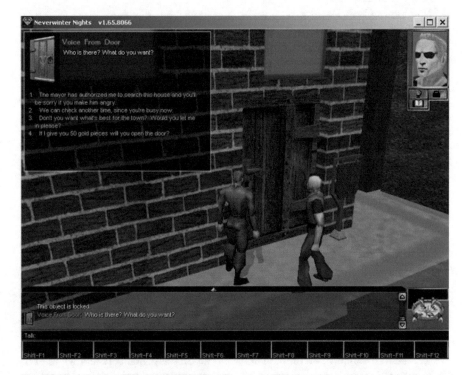

Fig. 7.2 A screenshot from the SABRE game based test-bed

each, were required to search for hidden weapons caches in an urban environment (town) while earning or loosing Goodwill points. Different amount of Goodwill points were earned depending on whether the weapons cache was found indoor or outdoor. Team also can lose points if for example they open a weapon-less container, etc. Players have a significant choice over the amount, timing, and type of interactions like chatting to specific individuals or broadcast, communication using structured formats using the journal-management or map-marking tools provided to the members. There were several phases in the game starting with Survey, followed by Training and Planning phases and finally, the Executing phase. It is the Execution phase, 1 h in length, where the four member teams search for the weapons and earn good will points.

7.2.2   Individual Level Metrics

In our analysis we develop two types of Individual Level Metrics from the SABRE dataset. The first are the **Role** type metrics. These are based upon the kind of role the individual is playing within the team. There are a total of seven Role Metrics for each individual member of a team:

1. Number of Tips from NPC (Non-Player Character--automated in the game)
2. Number of Conversations initiated with NPC
3. Number of Chats Sent
4. Number of Chats Received
5. Number of Buildings Entered
6. Number of Tips Sent
7. Number of Tips Received

These metrics try to quantify the Role an individual is playing within the team while keeping track of the various actions he or she performs or his/her in-game dynamics.

The second type of metrics are the **Skill** type metrics which reflect upon the skill of a team member. These were ascertained via a pre-game survey filled by each of the members for all the teams. In all we have 18 different kinds of Skill-type individual metrics (Table 7.2).

7.2.3   Constructing Group Level Metrics (Control Variables) from Individual Metrics

We now develop group or team level metrics using the two types of Individual Metrics discussed in the previous subsection. We construct the group level metrics by aggregating the individual level metrics for all the four individuals in each group. We aggregate in two ways to get two kinds of group level metrics. For the first kind, we take sum of values of an individual metric for all team members and we refer to these as the "TOTAL" group metrics. The second group metric is attained by taking

Table 7.2 List of skill type individual metrics with their type and range

Member Skill	Type	Value
English native	Yes or no	{1,2}
English ability	4 level choices	{1,2,3,4}
Stress in English environment	4 level choices	{1,2,3,4}
Reserve for English view	5 level choices	{1,2,3,4,5}
Computer expertise	3 level choices	{1,2,3}
Own computer	Yes or no	{1,2}
Email usage	5 level choices	{1,2,3,4,5}
Browser usage	5 level choices	{1,2,3,4,5}
Teleconference usage	5 level choices	{1,2,3,4,5}
Chat usage	5 level choices	{1,2,3,4,5}
Net-meeting usage	5 level choices	{1,2,3,4,5}
Own game console	4 category choices	{1,2,3,4}
Comp games time spent	Number of hours	Real
Multiplayer comp game	Yes or no	{1,2}
Neverwinter Nights	Yes or no	{1,2}
Comp game names	Yes or no	{1,2}
Game mods	Yes or no	{1,2}
Game list	Yes or no	{1,2}

into consideration the heterogeneity among the group members with respect to a given individual metric. We quantify this heterogeneity by employing the concept of Information Entropy (Teachman, 1980). We define the Information Entropy for a group of four members for a given individual metric "x" as:

$$H(x) = -\sum_{n=1}^{4}\left(p_n \log_2 p_n\right) \qquad (7.1)$$

where

$$p_n = \frac{x_n}{\sum_{n=1}^{4}(x_n)} \qquad (7.2)$$

is the fractional contribution of the member n for individual metric x and x_n is the value of the individual metric x for the member n of the group. As there are only four members in each group we have H in the range $[0,2]$. The higher the entropy, the lower the heterogeneity. Table 7.3, illustrates the values for the values attained by "TOTAL" and "ENTROPY" metrics for some example values of the "Tips Sent" individual metric i.e. x = "Tips Sent".

Tables 7.4 and 7.5, show the Group Level Metrics corresponding to the Role and Skill Type Individual Metrics, respectively, along with their mean values across all the 56 Groups in the SABRE dataset.

Table 7.3 Four example teams with different kinds of variety with respect to tips sending behavior. Tips Sent Entropy and Total metrics are also shown

Attribute: tips sent					
Member 1	Member 2	Member 3	Member 4	Entropy metric	Total metric
1 (p1 = 1/8)	0 (p2 = 0/8)	1 (p3 = 1/8)	6 (p3 = 6/8)	1.06	8
6	6	5	6	1.99	23
0	0	0	1	0	1
6	6	6	6	2	24

Table 7.4 List of all the group level role type metrics along with their mean values across groups

Total role metric	Mean value	Entropy role metric	Mean value
Tips_from_NPC_Total	17.625	Tips_from_NPC_Entropy	1.770445
NPC_Interacted_Total	85.98214	NPC_Interacted_Entropy	1.711167
Chats_Received_Total	657.1607	Chats_Received_Entropy	1.982355
Chats_Sent_Total	657.1607	Chats_Sent_Entropy	1.87555
Buildings_Entered_Total	61.33929	Buildings_Entered_Entropy	1.847229
Tips_Received_Total	23.96429	Tips_Received_Entropy	1.492368
Tips_Sent_Total	23.96429	Tips_Sent_Entropy	1.537586
Total_Mean_Total	218.1709	Total_Mean_Entropy	1.7452

Furthermore, we also have information per team regarding the type of configuration they adopted while playing the game. There are five group configurations as follows:

1. {1-1-1-1}: All working separate.
2. {1-1-2}: Two working together and the other two separately.
3. {1-3}: One working separately and three together.
4. {2-2}: Working in groups of two.
5. {4}: All working together.

Corresponding to the above five group configurations we have define five TOTAL Group Level Metrics:

1. Group_Conf_1-1-1-1_Total: Percentage of time spent in configuration {1-1-1-1} configuration
2. Group_Conf_1-1-2_Total: Percentage of time spent in configuration in {1-1-2} configuration
3. Group_Conf_1-3_Total: Percentage of time spent in configuration in {1-3} configuration
4. Group_Conf_2-2_Total: Percentage of time spent in configuration in {2-2} configuration
5. Group_Conf_4_Total: Percentage of time spent in configuration in {4} configuration

Table 7.5 List of all the group level skill type metrics along with their mean values across groups

TOTAL skill metrics	Mean	Min	Max	ENTROPY skill metrics	Mean	Min	Max
English_Native_Total	4.678571	4	8	English_Native_Entropy	1.986054	1.9219	2
English_Ability_Total	11	7	16	English_Ability_Entropy	1.967525	1.8911	2
Stress_in_English_Total	9.642857	6	14	Stress_in_English_Entropy	1.941323	1.5305	2
Reserve_for_English_View_Total	13.14286	10	18	Reserve_for_English_View_Entropy	1.962373	1.8352	2
Game_Mods_Total	4.142857	4	5	Game_Mods_Entropy	1.988843	1.9219	2
Comp_Game_Names_Total	3.178571	1	4	Comp_Game_Names_Entropy	1.612232	0	2
Neverwinter_Nights_Total	4.285714	4	6	Neverwinter_Nights_Entropy	1.980346	1.9183	2
Multiplayer_Comp_Game_Total	7.655357	0.2	27.2	Multiplayer_Comp_Game_Entropy	1.124491	0.10125	2
Game_List_Total	4.232143	4	6	Game_List_Entropy	1.9832	1.9183	2
Email_Usage_Total	19.44643	16	20	Email_Usage_Entropy	1.993227	1.8232	2
Browser_Usage_Total	19.08929	13	20	Browser_Usage_Entropy	1.980116	1.8232	2
Teleconference_Usage_Total	8.267857	4	15	Teleconference_Usage_Entropy	1.854605	1.65	2
Chat_Usage_Total	13.94643	5	20	Chat_Usage_Entropy	1.876964	1.65	2
Netmeeting_Usage_Total	6.482143	4	13	Netmeeting_Usage_Entropy	1.858314	1.5488	2
Own_Game_Console_Total	6.071429	4	11	Own_Game_Console_Entropy	1.928354	1.75	2
Comp_Games_Time_Spent_Total	15.54179	0.04	62.04	Comp_Games_Time_Spent_Entropy	1.147974	0.029089	2
Computer_Expertise_Total	9.482143	6	12	Computer_Expertise_Entropy	1.971993	1.8911	2
Own_Computer_Total	3.946429	3	4	Own_Computer_Entropy	1.977768	1.585	2

We also define one ENTROPY metric for group configuration which captures the diversity in group configuration over time. We refer to it as, "Group_Conf_Entropy".

7.2.4 Group Performance (Dependent Variables)

As the teams search for weapons they earn or lose goodwill points. We define Performance of a team as the Net Change in number of goodwill points earned by each team. The histogram of team performance is shown in Fig. 7.3. The middle of the three red vertical lines is the mean performance (**840.71**) and the other two denote the top and bottom 25 % performance cutoff for teams. We use these cutoffs to define three categories (0, 1 and 2) of team as follows:

Category 0—Low Performing teams (bottom 25 %): Net Goodwill points $\leq$500.
Category 1—Medium Performing teams: 500 < Net Goodwill points < 1150.
Category 2—High Performing teams (top 25 %): Net Goodwill points $\geq$ 1150.

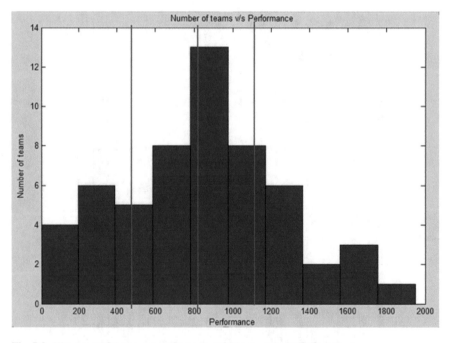

Fig. 7.3 Histogram of the group performance of 56 groups in SABRE dataset

7.3   Experimentation Methodology

Our experiments involve the application of machine learning methodologies described in Sect. 7.1 to perform group analysis of teams in the SABRE dataset. We divide the experiments into two types of major levels (see Fig. 7.4). First, is the Micro-Level analysis where we perform the group analysis using a single type of group metrics (variables). As we have three types (Role, Skill & Group Configuration) of group-level metrics, the Micro-Level contains three experiments where we only consider attributes from within each of these three types. Second, we have Macro-Level analysis where we consider all the three type of metrics simultaneously. Within the Macro-level we consider all the three metrics together.

As the reader can observe each of the just described experiments different in the type of group attributes employed for analysis. Each of these experiments is conducted in four phases (see Fig. 7.5). Each phase helps us understand, from a variety of perspectives, including insights from their attributes (features), their relationships, and their effects on the group performance. We start with simple correlation

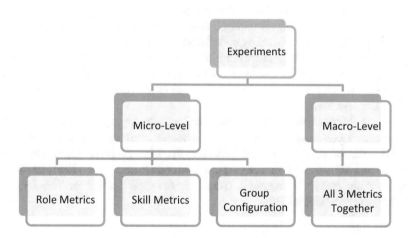

Fig. 7.4   Segregation of the different types of analysis conducted

Correlation Analysis	Decision Trees	Feature Selection	Decision Trees on Selected Group of Metrics
• How individual metrics affect the performance? • How pairs of individual metrics affect each other and the performance?	• How group of individual metrics affect each other and the performance?	• Select the most important group of metrics that affect performance?	• How group of individual metrics affect each other and the performance?

Fig. 7.5   Diagram showing the various analysis phases along with their purposes

analysis to find pair-wise dependence between all variables, both within and between each dependent and independent variable. This is followed by a decision tree, which explicitly highlights the patterns of relationships between different variables that may affect group performance. We perform feature selection next in order to focus on the dominating or most explanatory variables and discuss why the selected features can possibly be relevant. Finally, we again perform decision tree analysis using on the selected features from the previous phase and hope to find more strong and interesting patterns. We overall, therefore, have four sets of experiments and in each experiment we analyze groups from a series of four phases as we just described. Also within each of the four sets we consider both the TOTAL and ENTROPY variants of the group metrics.

7.4 Experiment 1: Group Analysis Using Role Based Metrics

7.4.1 Phase 1: Correlation Analysis

Table 7.6 shows the correlations with group performance among the different independent variables. The total amount of Tips Sent (total metric correlation of 0.43) and entropy of Tips Sent (entropy correlation of 0.30) were both significantly correlated. There was also a negative correlation with entropy regarding the number of buildings entered (negative entropy correlation of −0.22). Overall, it also seems that the TOTAL metrics are more related in general to the performance rather than the ENTROPY metrics.

The correlations between total group level metrics suggest some interesting and explainable dependencies (Table 7.7). For example, the more a team interacts with the NPCs the more likely the team gets more tips from them (correlation of 0.646). Also, as one of the team member gets tips from NPCs he or she is likely to forward them to other members, therefore, increasing the total tips flux within the group (observe the correlation 0.30).

Table 7.6 Correlation between independent variables and performance (dependent variable)

Total role metric	Correlation score	Entropy role metric	Correlation score
Tips from NPC	0.383145	Tips from NPC	−0.009487
NPC interacted	0.310966	NPC interacted	0.104403
Chats received	0.269815	Chats received	−0.102956
Chats sent	0.269815	Chats sent	**−0.221359**
Buildings entered	0.279464	Buildings entered	−0.221359
Tips received	0.430349	Tips received	0.081854
Tips sent	**0.430349**	Tips sent	**0.300998**
Total mean	0.339129	Total mean	0.0066

Table 7.7 Pair-wise correlation between total role type group metrics

Correlation	Got_Tip_from_NPC_Total	NPC_interacted_Total	Chats_Received_Total	Chats_Sent_Total	Buildings_Entered_Total	Tips_Received_Total	Tips_Sent_Total
Got_Tip_from_NPC_Total	1.000						
NPC_interacted_Total	**0.646**	1.000					
Chats_Received_Total	0.093	0.107	1.000				
Chats_Sent_Total	0.093	0.107	1.000	1.000			
Buildings_Entered_Total	0.187	0.128	−0.022	−0.022	1.000		
Tips_Received_Total	**0.300**	0.229	0.135	0.135	0.101	1.000	
Tips_Sent_Total	**0.300**	0.229	0.135	0.135	0.101	1.000	1.000

Let's focus now on the Entropy metrics and their pair-wise correlations, as depicted in Table 7.8. High Entropy for a given variable indicates that team members behave similarly with respect to that variable and Low Entropy indicates that there is a large variation among the team members for the given variable. Now we see a pretty high correlation between the entropies of interactions initiated with NPCs and the tips received by NPCs (correlation 0.595). This may make sense because if everyone initiates a conversation with NPCs (high entropy of initiation) everyone is likely to get a tip (high entropy of tips from NPC). Similarly, if only a few interact with NPCs (low entropy for initiation) only those few would receive tips from NPCs (low entropy). Although, this argument is straight forward, the point we wish to highlight is that this reasoning is not possible without a team diversity metric like entropy.

Further more interesting would be to utilize the correlation between the entropy metrics and the total metrics as shown in Table 7.9. For example, we observe a negative correlation between Chats received as well as the Chats sent entropy and the total amount of buildings entered by the team. A possible explanation would be that team is busy in chatting and therefore, fail to enter several buildings. Also chat-receiving entropy is negatively correlated with the total amount of tips received from NPC (correlation −0.226). This suggests that possibly a few team members are busy getting tips from NPC (making high total NPC tips for team) and these members are not receiving much chats, as compared to other members (low entropy), because they are busy interacting with NPCs.

7.4.2   Phase 2: Decision Tree Analysis

Weka was employed for Decision Tree Analysis using the J48 Decision Tree implementation provided in the software. To give a more hands-on experience, Fig. 7.6 shows the "Preprocess tab" when we load the data (only the Role type group metrics) in the Weka software.

In order to perform decision tree analysis we move to the "Classify" tab (see Fig. 7.7) and choose using the "Choose" button the J48 (which can be found under weka >classifier −> trees) classifier. Run the classifier using the "Start" button on the left after choosing the "Use training set" option under the "Test options".

At this point, we would again highlight here that our major focus in these experiments is not to build strong predictive models where the only concern is to improve the prediction accuracy over the unseen examples as a test set. Contrary to this, our main focus is to perform feature space analysis which involves objectives like reducing the number of independent variables to a manageable set. Furthermore, we would like to understand how the various features interact and which are the most important features that can help us understand the given data samples sufficiently well, rather than the generalization power of model to unknown test samples.

In other words, we are satisfied if our model fits the training data sufficiently well and focus on interpretation of feature space. For this reason, we choose the "Use training set" option under the "Test options" on the left. This tells Weka to evaluate the accuracy of the learnt model on the training data itself.

Table 7.8 Pair-wise correlation between role type entropy group metrics

Correlation	NPC_tips_ Entropy	NPC_initiated_ Entropy	Chat_Received_ Entropy	Chat_Sent_ Entropy	Buldings_Entered_ Entropy	Tips_Revc_ Entropy	Tips_Sent_ Entropy
NPC_tips_Entropy	1.000						
NPC_initiated_Entropy	**0.595**	1.000					
Chat_Received_Entropy	0.324	0.114	1.000				
Chat_Sent_Entropy	0.031	−0.010	0.196	1.000			
Buldings_Entered_Entropy	0.072	−0.075	0.064	0.004	1.000		
Tips_Revc_Entropy	−0.142	−0.140	−0.098	−0.001	0.099	1.000	
Tips_Sent_Entropy	−0.076	0.051	−0.024	0.012	−0.141	0.346	1.000

Table 7.9 Correlation between role type entropy v/s total group metrics

Entropy v/s Total	Got_Tip_from_NPC_Total	NPC_interacted_Total	Chats_Received_Total	Chats_Sent_Total	Buildings_Entered_Total	Tips_Received_Total	Tips_Sent_Total
NPC_tips_Entropy	−0.090	0.046	0.064	0.064	0.058	−0.097	−0.097
NPC_initiated_Entropy	−0.022	0.080	0.152	0.152	0.104	−0.063	−0.063
Chat_Received_Entropy	**−0.226**	−0.049	0.159	0.159	**−0.293**	−0.118	−0.118
Chat_Sent_Entropy	−0.018	0.014	0.324	0.324	**−0.197**	0.248	0.248
Buldings_Entered_Entropy	−0.226	−0.123	−0.068	−0.068	0.155	0.109	0.109
Tips_Revc_Entropy	−0.138	−0.116	−0.237	−0.237	0.158	0.369	0.369
Tips_Sent_Entropy	0.184	−0.031	0.300	0.300	0.005	0.567	0.567

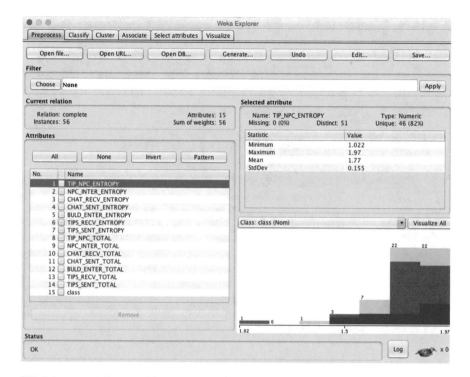

Fig. 7.6 Preprocess tab in Weka

After running the analysis, the output screen on right shows the results, as shown in Fig. 7.7. As we can observe the decision tree fits the 56 group samples fairly well. In order to visualize the tree, right click on the Result list at the bottom left and choose "Visualize Tree" option. Figure 7.8 shows the resultant tree for the total as well as entropy type group metrics together. Recall we had mentioned that we divide the teams in three categories: low (0), medium (1) and high (2), based upon their performance. Our aim in the Decision Tree analysis is to find those path ways or relationships between different variables starting from the *top* of tree that take us to high performing (labeled 2) *leaves* i.e. bottom-most nodes (dependent variable) in the tree. This helps us better understand the relationship in a visual fashion. Note the format of the leaves in the decision tree is of type x(y/z) where x is the class label (0: low, medium:1 or 2:high), y is the number of samples or instances correctly classified and z is the number of samples incorrectly classified. We would like to have the fraction (y/z) as high as possible for a reliable decision on the leaf node.

We observe in Fig. 7.8, that sub-tree to the right of the nodes: TIPS_RECV_ TOTAL and TIPS_SENT_ENTROPY, contains mostly medium and high performing leaves. Therefore, higher tips circulated within the team and higher tips sent entropy are all related to team performance according to the model (i.e. everyone sending tips results in good team performance).

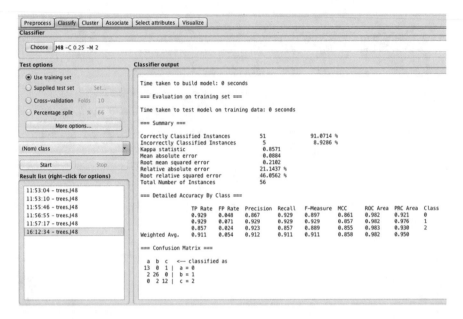

Fig. 7.7 Full role metric model fit statistics

Also, we observe that the high performing teams (leaves with label '2') are either in the right sub-trees of TIPS_SENT_ENTROPY node or of the NPC_INTER_ TOTAL node. But we can notice that even after this, if a group falls in the left sub-tree of NPC_INTER_TOTAL ($\leq$91) node, it is still predicted to have medium performance by having sufficiently high (>18) total tips from NPCs (i.e. right of TIP_NPC_TOTAL is a medium ('1') leaf). This reflects the importance of tips from NPCs. However, we also find three high performing groups (leaf labeled '2 (3.0)') to the left of TIP_NPC_ENTROPY. This means that if the tips receiving entropy of the group is less than 1.7 it is predicted to be high performing. For a four member team this typically should mean that only one or two members should be receiving those tips from NPCs. Readers are encouraged to see Table 7.3 to get a sense of the range of entropy and the type of values assumed by team members for a metric.

Note that we chose the minimum number of classified instances as two using the "−M" option for our classifier as "**J48** −C 0.25 −M 2" (see top of Fig. 7.7). This means that our decision tree will assign a new variable node even if the instances it is able to split are as low two. Therefore, if the leaf format in the visualized tree is x(y/z) then y $\geq$ m if we select option "−M m". In our case we observe this limit in the leaf "0(2.0/1.0)" where y = 2 as we chose m = 2. Notice that as we increase 'm', the misclassification instances i.e. z will also increase. We however, leverage the small size of our data to completely interpret our data by generating a tree node even if it is able to classify as low as two instances only.

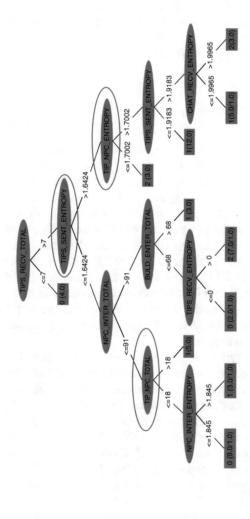

Fig. 7.8 Full role metric model decision tree

Adding to discussion on generalizability of the models we are using, we would also bring into notice that our dataset size is quite small and sparse with the sample size being the same as the number of dimension (56 sample size and 56 metrics with total and entropy types combined). Therefore, the generalizability and prediction on out of sample test cases for our models is not high. But they very well explain the training samples and how features affect the given data. We have chosen this smaller dataset in order to illustrate how beautifully we can zoom into the feature space. Our focus is therefore, how well can the group features explain the data samples. So in some sense we are fitting the machine learning model to the training set and care less about the prediction capability. If we have a larger dataset we can have more generalizability and less prediction error on testing this set as well.

7.4.3   Phase 3: Feature Selection

In the previous two sections, our analysis consisted of all the 16 available metrics of the role type. However, not all the metrics might be that relevant for a performance analysis of the teams. In machine learning, a subset of the most important variables and rank among them is done using feature selection methods (Guyon et al., 2010). Although there are a variety of feature selection methods, we will focus on of the powerful SVM classification based embedded method (Guyon et al., 2002) discussed earlier. In the Weka software this SVM based method is implemented under the name "SVMAttributeEval" in the Attribute Evaluators which is under "Select Attributes" tab (see Fig. 7.9). There are several options within SVMAttributeEval that we can play with, but for this illustration we restrict to the default options. Note, "attribute evaluator" scores the worth a subset of features and "search method" determines what kind of search is performed. We encourage readers to try different kinds feature selection methods.

After pressing the "start" button, the method returns a ranked list of all the attributes as per their relevance (as can be observed in the Attribute Selection output on the right). SVM-RFE algorithm implemented within "SVMAttributeEval" eliminates as well as rank the features iteratively. In each iteration the features are eliminated if required and are ranked as per their performance classification accuracy over the training set when used within the SVM classifier. We observe in the selected features, similar to the decision tree analysis in the previous section, that the tips exchanging behavior of members, captured in TIPS_SENT_ENTROPY and TIPS_SENT_TOTAL metrics, plays an important role in deciding team success. Furthermore, unlike any of the previous analysis, feature selection also indicates that chatting behavior of members also affects the performance.

Recall we are only concerned with the accuracy of the model on the training set and therefore, we choose the "Use training set" option under "Test options" on the left. If we have a larger sample size, then we can go for cross-validation as well. In fact, for our data, both for decision trees as well as feature selection, there was almost no difference between the models built using training set (with low error)

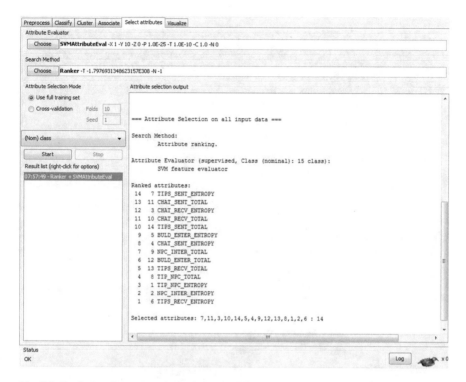

Fig. 7.9 Ranked attributes for role model using SVM

and via cross-validation (with less accuracy). This further confirms that our generalizability is restricted by lack of enough data samples. We therefore, focus on training set performance only.

7.4.4   Phase 4: Decision Tree Analysis over Selected Features

Notice that decision trees, as we saw in Phase 2, can tell us exactly whether it was the low or high value of a variable and in what context of other variable's values, affects group performance. This is in contrast to the black box approach of feature selection in Phase 3, which gives a list of highly important variables, but there is no way to ascertain what kind of values of these selected features affect the performance in what way.

In this phase we try to combine the best of both worlds. We use the top five highly ranked features, which in our case are the group level role metrics. In this way we leverage the ranking information from feature selection to lower the size of feature set from 16 to the five most important ones. We then build decision trees using only the top five role metrics just selected.

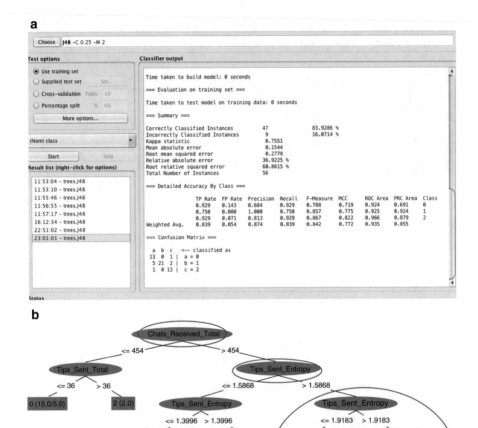

Fig. 7.10 (**a**) Reduced role metric model fit statistics. (**b**) Reduced role metric model decision tree

Before we go ahead with analysis of decision tree, we would like to comment on how to choose number of top ranked features. This choice is more of an art, especially if our focus is on feature space interpretation. Now as we increase the K top attributes, the training error on the decision tree built on it decreases. On the other hand, the number of attributes increases, making the tree possibly cumbersome to analyze. However, the latter is not always the case. Therefore, it becomes more of a subjective choice of K, which gives an interestingly interpretable decision tree and might have a sufficiently low training error as well.

For our choice of top five, the resultant tree is shown in Fig. 7.10b above and the model fit on training data is shown in Fig. 7.10a. As we can see that we now have a tree of just four metrics out of the 5 previously selected in Phase 3. This tree is sufficiently detailed and precisely tells us which kind of groups fall in intersection of which values of just these four group metrics. The big marked circle on the right contains a sub-tree whose leaves are either medium or high performing, implying that if a team falls in this sub-tree it is highly probable that it would perform well (at least medium if not high). In order to fall in this sub-tree, the team members should be chatting a lot and should have a similar tip sending behavior among the members (see the nodes in the two small circles).

Also if we observe the root node (CHAT_RECV_TOTAL), the left of root occurs if a group is chatting quite a bit less (<454). This value is significantly lower than the mean total chat across groups (see Table 7.4). If group members chat less and do not also send tips much, i.e. fall on left of TIPS_SENT_TOTAL node (left of root node), this group is more likely to perform low. As we can see the label of the leaf to the left of TIPS_SENT_TOTAL as "0(15.0/5.0)". There are 15 low performing groups out of the total 19 low performing groups that were predicted to fall in this leaf. Now, on the other hand, notice if we concentrate on the right of TIPS_SENT_ TOTAL. This happens if a low chatting group has significantly high (>36) tips circulated in group. Note the mean TIPS_SENT_TOTAL, from Table 7.4, is approximately equal to 24. So what this tells is that even a very low chatting group, if its members are circulating large volume of tips (> 36 much greater than mean of 24), it is predicted to perform well. As the label leaf on the right of TIPS_SENT_ TOTAL says there are only two such cases seen so far i.e. "2 (2.0)". Therefore, although such events are possible, they are very unlikely. So it is best for the team members to chat more (i.e., over 454).

Also if we observe the two TIPS_SENT_TOTAL nodes (one on top left and one bottom of the tree), we realize that higher total sent tips results into high performance even if the team is chatting less and has less tip sending entropy (i.e. only a few members send a large number of tips). As such, this indicates that high tip sending behavior may be favorable in the absence of chat receiving.

Summarizing this example, we found through four different types of analysis that for good performance, everyone in team should be communicating via both chatting as well as exchanging tips, but only a few members should be receiving a lots of tips from NPC and entering buildings.

7.5   Experiment 2: Group Analysis Using Skill Based Metrics

7.5.1   Phase 1: Correlation Analysis

We shall proceed for the group analysis using Skill metrics in a fashion similar to Role metrics performed in the last experiment. The Skill metric largely refers to the diversity of skills that make up each team and can be important regarding the assembly of teams. However, this time we assume that, with the detailed description in previous example, the reader is acquainted with the interpretation of entropy metrics as a variety quantifier. Firstly, we will see the correlation with the performance variable on different independent variables (both total and entropy metrics for all the skill type variables) as shown in Table 7.10. All the interesting correlations are highlighted using bold font. Overall, total English and Computer expertise as well as Native English speaking ability in the team are good predictors (positive corr. = 0.396 between total English native speaking ability of team with performance) of group performance. However, only "few" Native English speakers are better (negative corr. = −0.186 between Native English speaking Entropy and Performance). Teams having the most members with knowledge of Computer (positive corr. = **0.408** between Computer Expertise and performance) and spending time on Computer games (positive corr. = 0.334 between Comp. Games Time spent and performance) had a positive relationship with team performance.

The correlations between Entropy and Total skill metrics are shown in Table 7.11. The diagonal of this table is pretty important and interesting. All the interesting correlations are highlighted with bold font in Table 7.11. If a particular diagonal element is highly positive, it implies that the variable representing this row/column is high for all the individuals (high entropy) if total sum of all the team members for this variable is high (high total). On contrary if this diagonal element is highly negative, then it suggests that when the total group metric for this variable is high (high total metric), then only few (possible 1 or 2 in our four team member case) members are responsible or have high value for this variable (low entropy). Let us explain this with an example. Observe that Browser_Usage_Total is highly correlated with Browser_Usage_Entropy (positive corr. = 0.885 highlighted on the diagonal). This means that if the total browser usage in a team is high then the entropy with respect to browser usage in the team is also high. High entropy means that all the members of the team exhibit similar behavior. Given that team has high total browser usage, this indicates that all the team members are equally contributing to this high browser usage of the team. Note that it could have been possible that only a single member is responsible for all or most of the browser usage. If this would have been the case, this cell corresponding to Browser_Usage_Total and Browser_Usage_Entropy would have been dark green (i.e. highly negatively correlated). In fact such is the case for the pair of Neverwinter_Nights_Entropy and Neverwinter_Nights_Total, which is highly negatively correlated with a value of −0.949. This indicates that if the total team's score for playing Neverwinter Night is high, then it is highly likely, in our four team member case, that it was possibly just single member responsible for this score (very low entropy).

This diagonal element property that we just stressed is very important as it highlights the importance of the two group level metrics Total and Entropy. This fine grained description that we are able to achieve just at the level of correlation analysis, shows the value of these group level metrics.

Table 7.10 Correlation between independent variables and performance (dependent variable)

Correlation with performance	English native	English ability	Stress in English environment	Reserve for English view	Game mods	Comp game names	Neverwinter Nights	Multiplayer comp game	Game list	Email usage	Browser usage	Teleconference usage	Chat usage	Net meeting usage	Own game console	Comp games time spent	Computer expertise	Own computer
Total metric	**0.396**	**0.435**	0.115	−0.011	0.078	0.345	0.186	**0.320**	0.246	0.017	0.071	−0.189	0.333	−0.001	0.324	**0.334**	**0.408**	0.108
Entropy metric	**−0.186**	0.021	0.029	0.244	−0.078	0.352	−0.229	−0.332	−0.257	0.019	0.079	−0.245	0.053	−0.150	−0.214	**0.240**	**0.238**	0.108

Table 7.11 Correlation between skill type entropy v/s total group metrics

Entropy v/s total	English_ Native_ Total	English_ Ability_ Total	Stress_ in_ English_ Total	Reserve_ for_ English_ View_ Total	Game_ Mods_ Total	Comp_ Game_ Names_ Total	Never-winter_ Nights_ Total	Multiplayer_ Comp_ Game_Total	Game_ List_ Total	Email_ Usage_ Total	Browser_ Usage_ Total	Telecon-ference_ Usage_ Total	Chat_ Usage_ Total	Netmeeting_ Usage_Total	Own_ Game_ Console_ Total	Comp_ Games_ Time_ Spent_ Total	Computer_ Expertise_ Total	Own_ Computer_ Total
English_ Native_Entropy	−0.114	−0.040	−0.158	−0.014	0.057	0.046	−0.102	0.164	0.032	−0.074	0.089	−0.005	0.096	0.075	0.047	0.069	0.151	−0.111
English_ Ability_Entropy	0.301	0.385	−0.129	0.000	0.031	0.158	0.195	0.159	0.120	0.216	0.414	0.010	0.134	−0.127	0.352	0.165	0.297	0.312
Stress_in_ English_ Entropy	0.205	0.266	0.052	−0.268	−0.028	0.113	0.171	0.134	0.058	0.284	0.295	0.112	0.078	−0.140	0.046	0.113	0.144	0.121
Reserve_for_ English_View_ Entropy	0.288	0.269	−0.031	−0.143	0.117	0.091	0.245	0.116	0.182	0.032	0.397	0.127	0.259	−0.033	0.302	0.154	0.244	0.265
Game_Mods_ Entropy	0.017	−0.110	−0.120	−0.103	**−1.000**	−0.036	0.125	−0.099	−0.678	0.136	−0.111	0.077	−0.130	−0.133	−0.013	−0.145	−0.314	0.130
Comp_Game_ Names_Entropy	0.107	0.126	0.034	0.043	0.073	**0.982**	0.090	0.458	0.156	−0.067	0.091	0.121	0.368	0.170	0.203	0.471	0.482	0.047
Neverwinter_ Nights_Entropy	−0.546	−0.509	−0.104	−0.154	0.119	−0.129	**−0.949**	−0.446	0.024	−0.245	−0.104	−0.054	−0.037	0.081	−0.434	−0.471	−0.315	−0.137
Multiplayer_ Comp_Game_ Entropy	−0.136	−0.273	−0.247	−0.215	−0.189	−0.354	−0.158	−0.272	−0.104	0.194	−0.162	−0.188	−0.339	−0.272	−0.076	−0.285	−0.354	−0.013
Game_List_ Entropy	−0.293	−0.337	0.095	0.057	−0.660	−0.154	0.036	−0.066	**−0.965**	0.060	−0.056	0.098	−0.121	0.017	−0.105	−0.191	−0.413	0.068
Email_Usage_ Entropy	0.133	0.162	0.206	0.085	−0.274	0.051	0.126	0.175	−0.203	**0.687**	0.272	0.062	0.117	−0.161	0.246	0.120	−0.043	0.543
Browser_ Usage_Entropy	0.126	0.191	0.105	0.011	−0.018	0.122	0.040	−0.070	0.044	0.091	**0.885**	0.285	0.159	0.173	0.193	0.081	0.190	0.475

Teleconference_Usage_Entropy	−0.080	0.093	0.176	0.019	−0.046	−0.145	0.073	−0.003	−0.085	0.213	0.174	−0.062	−0.210	−0.027	−0.029	−0.066	0.078	−0.066
Chat_Usage_Entropy	−0.064	0.007	−0.060	0.047	0.006	0.176	−0.185	0.356	0.038	0.062	−0.040	0.009	**0.749**	0.163	0.087	0.251	0.120	0.073
Netmeeting_Usage_Entropy	0.006	0.108	−0.071	−0.049	−0.104	−0.181	−0.090	−0.078	0.069	0.297	−0.084	−0.339	−0.300	**−0.675**	0.146	−0.026	−0.257	−0.007
Own_Game_Console_Entropy	−0.071	−0.245	−0.199	−0.175	−0.203	−0.075	−0.064	−0.062	−0.096	−0.311	−0.025	0.233	0.026	0.176	−0.290	−0.110	−0.251	−0.110
Comp_Games_Time_Spent_Entropy	−0.071	−0.060	−0.006	0.147	0.050	0.485	0.148	0.278	−0.028	−0.128	−0.161	−0.159	0.106	−0.032	0.045	0.311	0.046	−0.274
Computer_Expertise_Entropy	0.210	0.280	−0.149	0.003	0.021	0.083	0.212	0.181	0.158	0.277	0.129	−0.271	−0.144	−0.325	0.274	0.139	0.261	0.069
Own_Computer_Entropy	0.123	0.170	0.238	0.101	−0.130	0.053	0.130	0.148	−0.052	0.358	0.537	0.255	0.288	0.116	0.057	0.184	0.237	**1.000**

7.5.2   Phase 2: Decision Tree Analysis

The decision tree using the Skill based metrics is shown in the Fig. 7.11a and the corresponding model accuracy on the training instances is shown in Fig. 7.11b. We know from Table 7.4 that the average Teleconference_Usage_Total across all the groups is around 8. The leaf right of the root node (Teleconference_Usage_Total) is attained if the group has very high (>13) Teleconference usage relative to the mean of 8. Unfortunately, this leaf is labeled "0 (3.0)", meaning three low performing groups have been observed with such high Teleconference usage. Here, Daily

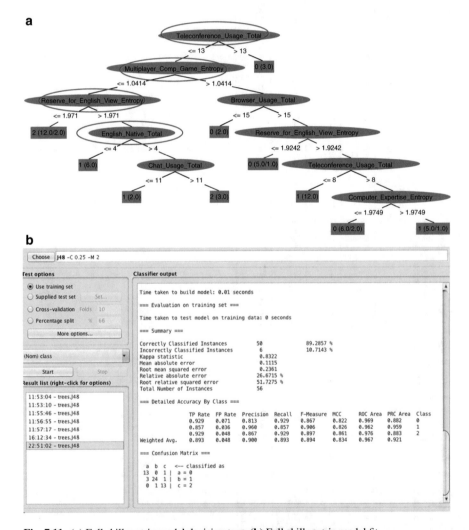

Fig. 7.11 (**a**) Full skill metric model decision tree. (**b**) Full skill metric model fit

or Weekly Teleconference usage was predicted as something agnostic to group performance.

Moreover, low entropy in multi-player game playing and low entropy in native English speakers had a positive relationship with high group performance. That is, groups with low entropy values on these two variables were overwhelmingly predicted to be in the high performance class. By following the parent nodes of these two variables (i.e. Multiplayer_Comp_Gam_Entropy ≥ Reserve_for_English_View_Entropy) to the leaves, it shows that 23 (e.g., add up all the predicted cases in the left, 12 + 6 + 2 + 3) groups fall in these leaves. Out of these 23 groups, 15 (12+3, ~65 %) were predicted as high performing groups (only two were incorrect). As such, out of the 14 high performing groups, these rules correctly classified 13 of them (~93 %), leaving only one false negative (i.e., a high performing group incorrectly predicted as not high performing).

7.5.3   Phase 3: Feature Selection

Similar to previous example, using Weka we performed the SVM based feature selection using the SVMAttributeEval functionality provided in Weka. The Attribute selection output contains the ranked list of various skill type group metrics is shown in Fig. 7.12.

7.5.4   Phase 4: Decision Tree Analysis over Selected Features

Finally, we perform a decision tree analysis using the selected features. This time we chose the top ten features out of the total 36 skill metrics which are shown in decreasing ranks in Table 7.12. We also tried other values for the number of top attributes to use, but they did not generate useful trees. In fact, the resulting J48 decision tree shown in the Fig. 7.13a employs only five features out of the ten selected features. However, as we can observe in Fig. 7.13a the left sub-tree of root has similar relationships to those in the tree built in Phase 2 (Fig. 7.10a). The left sub-tree highlighted with a red circle is the most interesting as it contains only high or medium performing groups. The corresponding model fit is given in the Classification output in Fig. 7.13b.

To summarize this example, the best predictor of high performing teams is a combination of low values regarding entropy in skill related to multiplayer computer games, total teleconference usage, and entropy in Reserve in English View presentation (12 predicted to be high performing, only two were incorrect). Likewise, high performing teams tended to have high entropy in Computer expertise and Game Mods (4 predicted to be high performing, one incorrect).

Fig. 7.12 Ranked attributes for skill model using SVM

7.6 Experiment 3: Group Analysis Using Group Configuration Metrics

In this section we focus on the effect of the group configuration metrics on the group's performance. Table 7.12 shows the correlation score of the different group configuration metrics with group performance (the dependent variable). The correlation of Group_Conf_1-1-1-1_Total with performance reflects that working separately is correlated with good performance, suggesting a division of labor may be beneficial rather than working collectively at the same time. To see this more visually we plot the linear regression curve in Fig. 7.14a where the line has a positive slope.

If we focus on the Group_Conf_Entropy, we observe a negative correlation with performance. Note that the Group_Conf_Entropy variable reflects homogeneity with respect to the different possible group configurations over time. It is high when a group spends equal time in each of the five configurations and lowest when the team is just playing in a single configuration during the entire playing time. The negative correlation therefore, suggests that in general, spending time in fewer

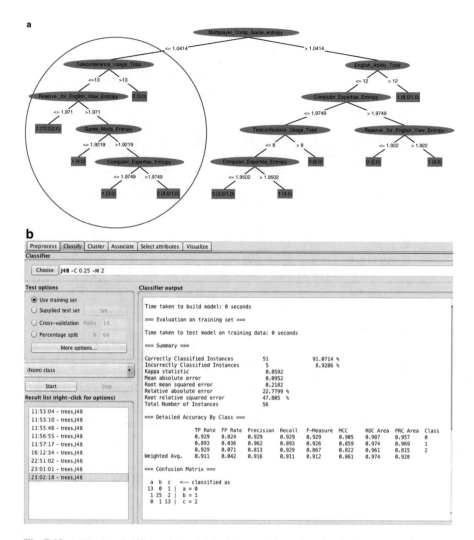

Fig. 7.13 (**a**) Reduced skill metric model decision tree. (**b**) Reduced skill metric model fit

Table 7.12 Correlation scores of the different group configuration metrics with group performance

Total metrics	Performance
Group_Conf_1-1-1-1_Total	**0.314**
Group_Conf_1-1-2_Total	−0.136
Group_Conf_1-3_Total	**−0.369**
Group_Conf_2-2_Total	−0.185
Group_Conf_4_Total	−0.177
Entropy metrics	
Group_Conf_Entropy_Entropy	**−0.349**

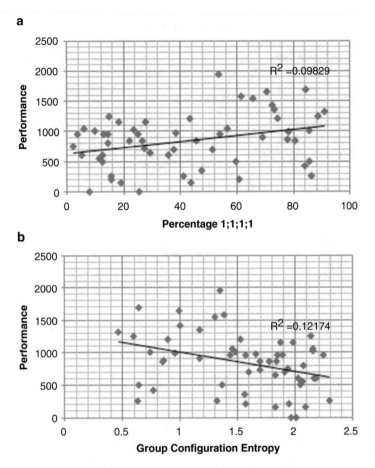

Fig. 7.14 (**a**) Plot of group performance and working separately. (**b**) Plot of group performance and Group Configuration Entropy

different configurations has a positive relationship with performance (rather than being equally distributed in all the configurations). This can be visually seen in the linear regression plot in Fig. 7.14b. Hence, analyzing both Entropy and Total metrics is important because both had significant relationships with group performance.

7.7   Experiment 4: Using All Types of Metrics Combined for Group Analysis

In this section we shall consider a mixed model that combines the set of all the three metric types: 16 role types, 36 skill types and 6 group configuration types together. Given we had already analyzed the correlation of these three types separately in the

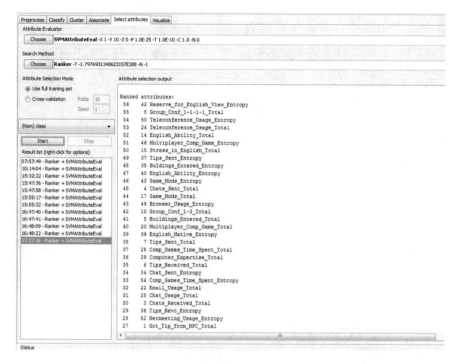

Fig. 7.15 Ranked attributes for mixed model using SVM

previous three experiments, we would not mention it further. We would also skip the pair-wise correlation analysis, although there are several interesting dependencies across different type metrics, but the across-type pairs are simply too many to analyze and describe. In fact, as the combined set of all the three types has a large number of metrics, we would directly perform decision tree analysis on SVM selected features (i.e. Phase 4). The attribute ranking of this combined set of metrics using the SVMAttributeEval in Weka is shown in the Fig. 7.15.

In this case we again go with top ten metrics, in decreasing order of rank. Moreover, the J48 decision tree classifier output tree is shown in Fig. 7.16b and the classification accuracy output from Weka is shown in Fig. 7.16a. The important point here is that now we are at the stage where we are considering all the 56 different metrics together. Therefore, we would now be able to compare and select metrics that are important across all the types. This should help us understand which are the most globally important metrics.

Although we had selected the top ten group metrics from the SVM ranking list, the decision tree only used eight out of these ten. We observe that among Role Metrics, Total Tips Sent, and Tips Entropy seem to play a very important role. That is, low performing groups tend to send few tips, while high performing groups tend to have higher levels of Total Tips and Tips Entropy. Within the Skill Metrics,

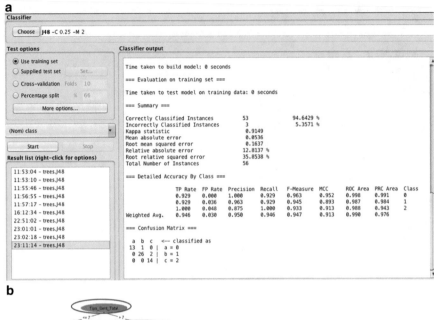

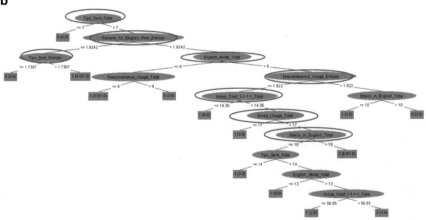

Fig. 7.16 (**a**) Mixed model fit. (**b**) Mixed model decision tree

heterogeneity of being Reserve in English View presentation and high English Ability tend to predict high performing groups. Teleconference Usage Entropy, Total Email, and Chat Usage turn out to be key factors as well.

As has been observed in previous analyses, total amount of time spent in {1,1,1,1} type Group Configuration is one the very crucial factors for team success. In fact, if we observe the pair-wise correlation matrix (see Table 7.13), we observe that when members work separately they chat less and spend more time in interaction with NPCs and gather tips from NPCs. This knowledge gathered from NPCs, we can hypothesize, may be highly influential for group success.

Table 7.13 Pair-wise correlation for mixed model variables

Correlation between total metrics	Got tip from NPC	NPC interacted	Chats received	Chats sent	Buildings entered	Tips received	Tips sent	Group_Conf_1-1-1-1_Total
Got tip from NPC	1.000							
NPC interacted	0.646	1.000						
Chats received	0.093	0.107	1.000					
Chats sent	0.093	0.107	1.000	1.000				
Buildings entered	0.187	0.128	-0.022	-0.022	1.000			
Tips received	0.300	0.229	0.135	0.135	0.101	1.000		
Tips sent	0.300	0.229	0.135	0.135	0.101	1.000	1.000	
1-1-1-1	**0.330**	**0.420**	**-0.344**	**-0.344**	0.243	0.071	0.071	1.000

Overall, the mixed model, beyond just being the best in terms of model fit, demonstrates how complex the interactions are amongst the different sets of variables. Following the different paths along the decision tree can yield important insights into how these variables moderate one another. As to which model is best depends on the goals of the researcher. All the models ran had overall accuracy levels nearing 90 %. As such, a parsimonious model, though slightly less accurate, may be useful for those attempting to seek out which are the "big" factors discriminating between high and low performing teams. On the other hand, a more complex, less parsimonious predictive model may be useful if the goal is to "predict at all costs", which may be useful developing predictive applications (e.g., a team assembly application).

7.8   Conclusion

In this work we illustrated how to analyze small group behavior using individual level data. In this direction we show two possible ways of aggregating individual level information to generate group level metrics. Further, we show how traditional correlation analysis can substantially be supplemented with the help of the proposed metrics. In this sense, the techniques are not competing, but complementary. Finally, we employ these metrics within existing machine learning and data-mining techniques and illustrate, with the help of Weka data-mining software, how group performance can be analyzed using data-mining.

References

Chandrashekar, G., & Sahin, F. (2014). A survey on feature selection methods. *Computers & Electrical Engineering, 40*(1), 16–28.

Guyon, I., & Elisseeff, A. (2003). An introduction to variable and feature selection. *Journal of Machine Learning Research, 3*(Mar), 1157–1182.

Guyon, I., Saffari, A., Dror, G., & Cawley, G. (2010). Model selection: Beyond the bayesian/frequentist divide. *Journal of Machine Learning Research, 11*(Jan), 61–87.

Guyon, I., Weston, J., Barnhill, S., & Vapnik, V. (2002). Gene selection for cancer classification using support vector machines. *Machine Learning, 46*(1–3), 389–422.

Hall, M., Frank, E., Holmes, G., Pfahringer, B., Reutemann, P., & Witten, I. H. (2009). The WEKA data mining software: An update. *ACM SIGKDD Explorations Newsletter, 11*(1), 10–18.

Quinlan, J. R. (1986). Induction of decision trees. *Machine Learning, 1*(1), 81–106.

Quinlan, J. R. (1993). *C4. 5: Programs for machine learning.* San Mateo: Morgan Kaufmann.

Teachman, J. D. (1980). Analysis of population diversity measures of qualitative variation. *Sociological Methods & Research, 8*(3), 341–362.

Warren, R., Diller, D., Leung, A., Ferguson, W., Sutton, J.L (2005). *Simulating scenarios for research on culture and cognition using a commercial role-playing game.* Proceedings of the 2005 Winter Simulation Conference, Orlando, FL, December 4-7, 2005.

Chapter 8
Simulation and Virtual Experimentation: Grounding with Empirical Data

Deanna Kennedy and Sara McComb

8.1 Introduction

A decade ago, Davis, Harrison, and their colleagues encouraged researchers in the organizational sciences to embrace simulation as a means of augmenting theory building in the field (Davis, Eisenhardt, & Bingham, 2007; Harrison, Lin, Carroll, & Carley, 2007). This call for simulation research in the organizational context was not the first (e.g., McGrath, 1981), nor the last (e.g., Wang, Zhou, & Zhang, 2016). It did appear, however, at a time when the computational tools available and researcher sophistication vis-à-vis computational approaches were aligned to stimulate the movement.

Simulation, in conjunction with virtual experimentation, is a useful tool for the organizational researcher because it facilitates comparative analyses of complex, multilevel team processes (e.g., Kozlowski, Chao, Grand, & Braun, 2016) that occur across a range of different contexts (Carley & Prietula, 1994; Davis et al., 2007; Prietula, Carley, & Gasser, 1998) and over time (Kozlowski, Chao, Grand, Braun, & Kuljanin, 2013). Results may be leveraged to inform theory building (Kozlowski et al., 2013) and guide the efficient design of future field or laboratory research, particularly when large resource expenditures may be required (Kennedy & McComb, 2014; Kennedy, McComb, & Vozdolska, 2011). Team researchers have embraced this simulation movement by examining, for instance, collective learning (Anderson & Lewis, 2013), interaction modes (Bhuiyan, Gerwin, & Thomson, 2004), cognition (Grand, Braun, Kuljanin, Kozlowski, & Chao, 2016), communication patterns

D. Kennedy (✉)
School of Business, University of Washington Bothell, Bothell, WA, USA
e-mail: deannak@uw.edu

S. McComb
School of Nursing and School of Industrial Engineering, Purdue University,
West Lafayette, IN, USA
e-mail: sara@purdue.edu

© Springer International Publishing AG 2017
A. Pilny, M.S. Poole (eds.), *Group Processes*, Computational Social Sciences,
DOI 10.1007/978-3-319-48941-4_8

(Kennedy & McComb, 2014), project complexity (Kennedy et al., 2011), communication frequency (Patrashkova & McComb, 2004), transactive memory (Ren, Carley, & Argote, 2006), and team member replacement (Solow, Vairaktarakis, Piderit, & Tsai, 2002).

McGrath (1981) suggests that all research designs pose dilemmas for the researcher in that tradeoffs must be made between *generalizability* to populations of interest, *precision* in the measurement and control of variables, and *realism* related to the context in which the behaviors would be observed. In characterizing a variety of research strategies (e.g., laboratory experiments, field studies, judgment tasks), McGrath depicts computer simulation as a viable theoretical (vs. empirical) approach to conduct unobtrusive research (i.e., no observation of behavior is required) of a particular behavior system. As such, simulation provides a solution that attempts to address generalizability and realism, at the expense of precision. When selecting this approach, the researcher consciously decides to accept this compromise, since no design can maximize all three aspects simultaneously.

Once the decision to use computer simulation has been made, researchers must decide what type of simulation procedures they want to develop and validate. For instance, some of the aforementioned examples from the team domain developed simulation procedures via mathematical interpretations of theoretical relationships (e.g., Anderson & Lewis, 2013; Grand et al., 2016; Patrashkova & McComb, 2004; Solow et al., 2002), whereas others grounded their procedures with empirical data (e.g., Bhuiyan et al., 2004; Kennedy & McComb, 2014; Kennedy et al., 2011). Neither approach is right or wrong. As with all research, determining the best path forward must be based on criteria such as the research questions of interest, access to data, availability of adequate theory to model, etc. Regardless of the approach selected, ensuring that the simulation procedures, when executed, provide results that depict a reasonable representation of reality is of primary importance.

Herein, our purpose is to demonstrate (1) how simulation procedures can be developed and validated with existing empirical data and (2) how these procedures can be executed to conduct virtual experiments. To accomplish this purpose, we demonstrate how empirically collected data can inform simulation procedures to answer what-if research questions; the answers to which can, in turn, guide future empirical data collection. We discuss two examples to demonstrate the range of what-if questions that may be addressed via this approach. First, we provide guidance for developing simulation procedures that incorporate continuous data. At the end of the chapter, we describe how this approach was implemented in Kennedy et al. (2011). In that research, we used available team-level, continuous, cross-sectional data that had been collected via questionnaires to examine *how* project complexity impacts the curvilinear relationship between team communication and performance identified in Patrashkova-Vozdolska, McComb, Green, and Compton (2003). Virtual experiments were conducted by executing the simulation procedures under varying levels of project complexity to garner insights about the communication-performance relationship.

Second, we demonstrate how to develop simulation procedures with discrete data. We then provide the example of this approach, from Kennedy and McComb

(2014), where we used transcribed and coded communication strings (i.e., discrete, longitudinal data) from a laboratory study to understand the relationship between team performance and *when* teams shift their conversations among different processes. Virtual experiments were conducted to ascertain *what* happens if certain process shifts occur earlier or later in the team's life cycle. Results of both studies inform theory about team communication and can be tested through laboratory and/ or field experimentation.

8.2 Basic Overview of Simulation and Virtual Experimentation

In the following tutorial, we will demonstrate two methods: simulation and virtual experimentation. Some researchers suggest that simulation is virtual experimentation (Davis et al., 2007). We forward the notion that they are two distinctly different methods that may be used together. Simulation requires computer code and random numbers. The computer code imitates processes in the real world and the random numbers represent the variability inherent in those processes. A simulation occurs when the computer code is executed or run.

Virtual experiments are not unlike laboratory experiments. Researchers use closed systems (computer code in virtual experiments or laboratory apparatus in laboratory experiments) and manipulate parameters within those systems to study how the parameters influence outcomes. In virtual experimentation, the experiment is conducted using simulation. Specifically, the computer code is executed multiple times to conduct one simulation run for each manipulation designed into the experiment. One benefit of virtual experimentation is that each manipulation can be run hundreds or thousands of times in a matter of minutes. Sample sizes this large in a laboratory would probably be cost prohibitive and require significantly more time. We are by no means suggesting that virtual experimentation should replace laboratory experimentation. Both types of experiments have a role to play in organization science. Laboratory experiments may provide data that can inform the computer code used for virtual experiments. The results from virtual experiments may help researchers select more meaningful manipulations for laboratory experiments.

8.2.1 Tutorial

A number of resources provide guidelines for conducting simulation in general (e.g., Law & Kelton, 2000) and simulating organizational phenomena in particular (Burton & Obel, 2011; Hulin & Ilgen, 2000; Larson, 2012). We expand on these guidelines to provide step-by-step instructions for conducting simulation and virtual experimentation research. Herein, we summarize the purpose each step serves, the actions required for each step, and the outputs a researcher can anticipate upon

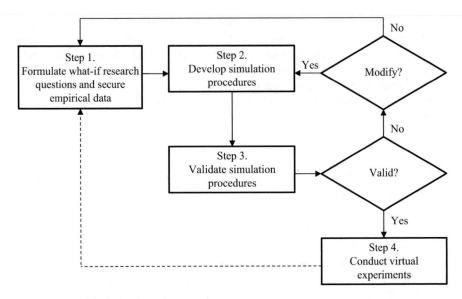

Fig. 8.1 Flowchart of key simulation steps

completion of the step. To depict how research progresses through these steps, we built a flowchart (see Fig. 8.1). As you can see in Fig. 8.1, ensuring that you have a valid model in the third step is an integral part of the process. If the simulation procedures are not valid, you have two options. On the one hand, if the results indicate that some minor tweaks might help you achieve validity, you may return to the second step and hone your simulation procedures. On the other hand, you may decide that validity is not attainable and go back to the first step to reformulate the research questions. The most important point here is that without a valid model, you should not move on to the fourth step and run virtual experiments.

In the following sections, we will expand upon this information by detailing the decisions and actions required for researchers to achieve their research objectives. Two basic examples, one with continuous data and the other with discrete data, will be used to demonstrate our points. These examples focus on the relationships among communication modalities, cohesion, and team effectiveness. Actionable computer code from Matlab R2015b is provided for these basic examples.

Step 1. Formulate what-if research questions and secure empirical data

Purpose: Establish coherence among your research questions, variables, and available data
Actions: Determine research questions regarding process of interest, identify corresponding variables, and secure empirical data
Outputs: Study goals, process parameters, and empirical data

This step might be called the chicken and egg step, because we don't always know if we will start with research ideas or the data required to test them when conducting simulation research. You may decide that you would like to expand on some results you obtained from a questionnaire field study, but your sample size is too small to ensure adequate power to test more complex models. Alternatively, you may have an interesting idea but inadequate resources to fully test it in a laboratory or field setting. In either case, you can formulate what-if research questions that can be addressed by simulation and virtual experimentation.

We use the scenario where data came first to develop our basic example using continuous data. Suppose we have data from an employee survey designed to solicit perceptions about the emergent state of team cohesion and team effectiveness from teams that use multiple communication modalities (e.g., face-to-face, computer-mediated). Our results support media synchronicity theory (Dennis, Fuller, & Valacich, 2008), which suggests a positive relationship between cohesion and effectiveness that is moderated by modality. The exact nature of that relationship within each modality cannot be tested though, because of our sample size. We are not sure that spending additional resources to collect more data will be a good investment. In this case, simulation can help us make the most of our empirical data and provide a more nuanced perspective of the cohesion-effectiveness relationship. The results from virtual experiments can help us decide if further laboratory or field research is warranted. Our what-if research question is: what-if teams use communication modalities with more or less synchronicity; does the cohesion-effectiveness relationship change? The variables of interest are cohesion, effectiveness, and modality. Empirical data from the employee survey is available and can be partitioned by communication modality.

We develop our discrete data basic example to demonstrate a scenario where an idea came first. While conducting a laboratory experiment requiring face-to-face teams to complete a planning task, we observed differences in the cohesion among team members. We also know that researchers have suggested that cohesion emerges and becomes pertinent as the teams complete their task (Kozlowski & Chao, 2012). Cohesion was not a variable of interest in the primary study, but we have access to a video of every team's laboratory session. If we develop a dataset that indicates when cohesion surfaces for each team, we can use virtual experimentation to examine the plethora of possible timing scenarios and compare the effectiveness achieved for each scenario. Such results may provide initial insights for theory building and provide information about the scenarios with the most potential for impactful results that can be tested in the laboratory or the field. In this example, our what-if research question is: what-if cohesion surfaces earlier or later; does the timing impact the cohesion-effectiveness relationship? The variables of interest are the point in time when cohesion emerges during the team's life cycle and team performance. By coding the videos, we can build the empirical dataset needed.

Step 2. Develop simulation procedures

Purpose: Articulate relationships among variables and the processes connecting them, ensure adequate power for generalizability, and convert simulation steps into actionable computer code

Actions: Identify simulation steps, establish stopping criteria, and code simulation procedures

Outputs: Logical depiction of the real-world process of interest, sample size requirements, and actionable computer code

Simulation procedures are commonly developed using the Monte Carlo simulation method. This method comes from a class of computational algorithms that uses repeated random sampling to generate results (Metropolis & Ulam, 1949). The method is implemented by drawing random samples from defined distributions and analyzing the behavior of the generated samples. Monte Carlo simulation has been applied in many different ways across organizational and management science, including imputing missing values (e.g., Newman, 2003; Roth, Switzer, & Switzer, 1999), estimating multistage outcomes (e.g., Charnes & Shenoy, 2004), and observing parameter variations (e.g., Avramidis & L'Ecuyer, 2006; Chance, Hillebrand, & Hillard, 2008; Valenzuela & Mazumbar, 2003). Building on this foundation, we will demonstrate how to apply random sampling to data distributions to (1) replicate the variables in the empirical dataset and (2) generate simulated datasets under changed or new conditions.

Before describing the specifics about how to develop simulation procedures, we need to discuss briefly when to stop a simulation run. Many approaches can be considered (e.g., convergence of the population (Georgieva & Jordanov, 2009)), but the common approach is to use sample size. Law and Kelton (2000), the seminal source for simulation, recommend a sample size of 10,000 to ensure robust sampling from the data distribution. Now, back to developing simulation procedures.

Step 2A. Identify simulation steps. The procedures you create should reflect your decisions about what variables need to appear in your simulated datasets to test your what-if research questions. At the same time, you need to ensure that you can replicate the way in which these variables connect in the real world. Both of our basic examples focus on the role of team cohesion; but the procedures required to generate the simulated data needed to test the what-if research questions may be quite different. These differences can be seen in Fig. 8.2. The diagrams in Fig. 8.2 are flowcharts (one for each of the cohesion-effectiveness examples) that identify what needs to happen in a computer program to build the desired simulated dataset. The flowcharts contain steps associated with initializing the simulation, generating and storing data, and determining if the simulation run should stop.

Step 2B. Code your simulation. Based on the flowchart you (or a valued collaborator familiar with computer coding) can code your simulation in a computer program that will provide efficient and reliable results. We have utilized Microsoft Excel, R, and Matlab for our own simulations. The way the diagram is translated into actionable code will depend on the coding language and the programmer's familiarity with the program. In essence you will want to think about how your simulation procedures play out mathematically.

Coding the simulation procedures for the continuous data basic example. To address the what-if research question in this basic example, we are interested in determining the relationship between cohesion and effectiveness under different scenarios of communication modalities (e.g., majority face-to-face or majority computer-mediated). Since cohesion and effectiveness are related, we can use the bivariate relationship to

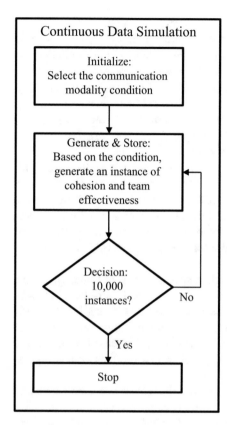

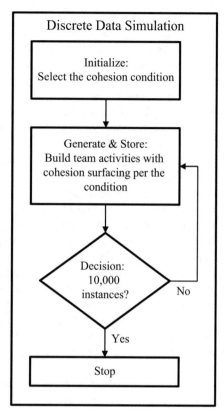

Fig. 8.2 Potential cohesion-effectiveness simulation diagrams

describe cohesion and effectiveness variables in the computer code. Assuming that the variables are normally distributed, we can represent them as continuous distributions with a mean μ and covariance Σ. The mean vector contains the mean values for cohesion and team effectiveness from empirical data, which is set at $\mu = (2.0, 8.0)$ for illustration. The covariance matrix includes the conditional distribution of team effectiveness on the distribution of cohesion and is represented as:

$$\Sigma = \begin{bmatrix} \pounds_{11} & \pounds_{12} \\ \pounds_{21} & \pounds_{22} \end{bmatrix} = \begin{bmatrix} 2.0 & 1.5 \\ 1.5 & 3.0 \end{bmatrix}$$

where $\Sigma_{12} = \Sigma_{21} = 1.5$ and indicates the covariance of cohesion and team effectiveness.

Now we are ready to simulate data by randomly drawing values for each variable from the bivariate distribution. Table 8.1 shows the translation of the survey study diagram into Matlab code for replicating the empirical dataset.

Coding the simulation procedures for the discrete data basic example. In this basic example, the what-if research question directs us to determine how the timing of cohesion emergence impacts effectiveness. To answer this question, longitudinal, discrete data about the team's state is needed. For illustration's sake, let us assume that we observed the team doing one of four possible activities at each minute during their 60-minute laboratory session. These observations were then converted into strings of activities such as Planning-Planning-Conflict-Conflict-Action-Cohesion-Cohesion-Action-Action ... and then transformed into the numerical string 4, 4, 3, 3, 1, 2, 2, 1, 1..., where Action = 1, Cohesion = 2, Conflict = 3, and Planning = 4. The numerical string contains not only the information about cohesion, but also information about when the other activities were being completed.

As researchers, we can make a decision to create computer code that either generates the entire string of activities or generates one number that represents where cohesion emerges. Either approach is acceptable. Generating the complete string requires more time to code and run the simulation, but may be useful for additional research. Alternatively, generating one number may be more expedient, but the actionable computer code will only be useful for this specific study. Our decision is to generate the complete string.

To start the process of generating complete strings, we have to determine the appropriate distribution(s) that best represents the empirical data. For this basic example, we assume two patterns were observed. First, as Gersick (1988, 1989) predicts, the distribution of activities is different before and after the midpoint of team activity. Second, the length of time the team spends focused on a specific activity is dependent upon the type of activity, but otherwise is reasonably consistent over time.

Let us assume that during the first 30 minutes, the teams' activities occur in a non-parametric manner. In this case, we can use the empirical data more explicitly to inform activity insertion into the data string. Table 8.2 shows the empirical distributions in terms of the probability mass function (pmf) and the cumulative distribution function (cdf). To determine which activity to insert during the first 30 minutes, (1) a random number between 0 and 1 is drawn from a uniform distribution, (2) the number is then matched with the appropriate interval of the cumulative distribution function (see top section of Table 8.2), and (3) the corresponding activity can be specified for insertion. Using the distribution information provided in Table 8.2, we can work through an example. If the random number generated for minute 17 was 0.624, the interval would be 0.501–1.000, the specific activity for minute 17 would be planning, and the number 4 would be selected for insertion into the numerical string of activities.

During minutes 31–60, the teams are observed engaging in all four activities in approximately equal numbers. Here, we can use a uniform distribution to select which activity to insert with each activity occurring about 25 % of the time (i.e., $U(1, 4)$). This scenario can simply be coded as drawing a random number from a discrete uniform distribution (i.e., =randi([1 4])) and inserting it directly into the numerical string of activities.

Once we know the activity to insert, the second step is to figure out how long the team does that activity. An initial investigation of the empirical data shows that each activity follows a reasonably uniform distribution pattern across minutes 1–30 and

Table 8.1 Continuous data simulation: procedures, validation, virtual experimentation

```
% Develop simulation procedures%
base_mu = [2, 8]; base_sigma = [2, 1.5;1.5, 3];
% Data simulation%
base_sim = mvnrnd(base_mu, base_sigma, 10000);
% Data storage%
dlmwrite('simulated survey data.txt', base_sim);

%Validation step%
display('Baseline model: RSquared = 0.38; F = 6076.40; p = 0.00; se = 1.90;')
base_betas = [6.49; 0.76]; baseCI = [6.45, 6.54; 0.71, 0.83]; base_pval = [0;
0];
rownames= {'Bo'; 'B1'}; BaselineResults = table(base_betas, base_pval,
baseCI, 'Rownames', rownames)

lm = fitlm(base_sim(:, 1), base_sim(:, 2), 'linear');
validation_betas = lm.Coefficients.Estimate; validation_pval =
lm.Coefficients.pValue;
ValidationResults = table(validation_betas, validation_pval, 'Rownames',
rownames)
if validation_betas(2, 1) >= baseCI(2, 1) && validation_betas(2, 1) <=
baseCI(2,2)
    display('Validation a success')
else
    display('Validation error, check and retry')
end

%Virtual Experimentation%
%high synchronicity condition = hs%
%low synchronicity condition = ls%
hs_mu = [4.5, 8]; hs_sigma = [2.4, 2.5;2.5, 4.2];
hs_sim = mvnrnd(hs_mu, hs_sigma, 10000);
display('Condition 1: High Synchronicity Simulated Dataset Generated')
ls_mu = [2, 5]; ls_sigma = [2.1,1.7;1.7,3.2];
ls_sim = mvnrnd(ls_mu, ls_sigma, 10000);
display('Condition 2: Low Synchronicity Simulated Dataset Generated')

%Analysis step
display('Condition 1: High Synchronicity Analysis Completed')
hs_lm = fitlm(hs_sim(:, 1), hs_sim(:, 2), 'linear');
hs_betas = hs_lm.Coefficients.Estimate;
display('Condition 2: Low Synchronicity Analysis Completed')
ls_lm = fitlm(ls_sim(:, 1), ls_sim(:, 2), 'linear');
ls_betas = ls_lm.Coefficients.Estimate;

%show together%
display('Plotting Regression Results across Experiments')
input = [ones(10000, 1) sort(base_sim(:, 1))];
output_base = input*base_betas; output_hs = input*hs_betas;
output_ls = input*ls_betas;
plot(input(:,2), output_base, 'r', input(:,2), output_hs,'b-.', input(:,2),
output_ls, 'g--')
legend('baseline', 'high synchronicity', 'low synchronicity')
```

a different, but still uniform distribution pattern occurs across minutes 31–60. As such, we can use a uniform distribution to represent the length of time spent doing the activity. The uniform distribution for each activity across time can be found in the bottom portion of Table 8.2. Continuing our example, we know that in the 17th minute, the team starts planning. We draw a random number between 0 and 8 (i.e., $U(0, 8)$) to determine how many minutes they spend on planning. If we draw a 3,

Table 8.2 The Distribution information for the discrete data simulation

Selection of *Which Activity* to Insert During Minutes 1–30				
X	1 (Action)	2 (Cohesion)	3 (Conflict)	4 (Planning)
pmf	0.125	0.250	0.125	0.500
cdf	0.125	0.375	0.500	1.000
interval	0.000–0.125	0.126–0.375	0.376–0.500	0.501–1.000
Selection of *How Many* Activity Observations to Insert				
X	1 (Action)	2 (Cohesion)	3 (Conflict)	4 (Planning)
From 1 to30 min	$U(1, 4)$	$U(2, 4)$	$U(2, 8)$	$U(0, 8)$
From 31 to 60 min	$U(2, 6)$	$U(1, 4)$	$U(0, 7)$	$U(1, 6)$

then minutes 17, 18, and 19 would be coded as planning (i.e., the number 4) in this activity string. We would then draw the next random number to determine the next activity that would start at minute 20.

With the distributions of variables in hand the coding of the simulation can proceed. Table 8.3 contains the Matlab code for this basic example. To represent the team's activities across 60 minutes, we will need to build the activity string for one team at a time by following an iterative cycle of selecting an activity and how many observations to assign that activity. Because the distributions of selecting an activity and the number of observations change at the midpoint, the code contains two loops. One decision we will have to make is how to handle the changeover in activities from the first 30 minutes to the second 30 minutes. As you will see in Table 8.3, if an activity is to be inserted across the 30-minute mark, it is carried over from the earlier period rather than truncated. We had to make this decision and could have as easily decided to truncate at the 30-minute mark. Such decisions are common in the translation of simulation procedures into computer code and must be made based on an amalgamation of research evidence, common sense, and practicality within the computer program selected.

Once the team's activity string is complete at a string length of 60, it is stored in a dataset, the variables are cleared, and a new team's activity string is initialized. Once the desired number of teams is reached, in this case 10,000 team activity strings, the program stops and the dataset is written to an external text file.

Step 3. Validate simulation procedures

Purpose: Compare simulation results with real-world data to ensure that the results reasonably represent reality
Actions: Test results from executing the computer code
Output: Valid, actionable computer code

Validation of your simulation procedures helps you ensure that your results are representative of the real world (Law & Kelton, 2000; Larson, 2012). In addition to adding rigor to your research process, this step also provides you with an opportunity to improve your simulation procedures, particularly with respect to the constructs needed to answer your what-if research questions.

Table 8.3 Discrete data simulation: procedures

```
% Develop simulation procedures%
storeteams = zeros(60, 10000);
selectobs = 0;
selectact = 0;
  for i = 1:10000;
  teaminst = []; m = 1;
  %for first 30 minutes%
  while m < 30;
    %Selects Activity 1, 2, 3, 4%
    selectless30 = rand();
    if selectless30 <= 0.125
        selectact = 1;
    elseif selectless30 <= 0.375
        selectact = 2;
    elseif selectless30 <= 0.500
        selectact = 3;
    elseif selectless30 <= 1.000
        selectact = 4;
    end
    %Selects Activity observation length%
    if selectact == 1
        selectobs = randi([1 4]);
    elseif selectact == 2;
        selectobs = randi([2 4]);
    elseif selectact == 3;
        selectobs = randi([2 8]);
    elseif selectact == 4;
        selectobs = randi([0 8]);
    end
    %Fills in the team instance string%
    teaminst(m:m+selectobs, 1) = selectact;
    m = m+selectobs;
  end
  %for second 30 minutes%
  while m < 60
  %Selects Activity 1, 2, 3, 4%
  selectact = randi([1 4]);
    %Selects Activity observation length%
    if selectact == 1;
        selectobs = randi([2 6]);
    elseif selectact == 2;
        selectobs = randi([1 4]);
    elseif selectact == 3;
        selectobs = randi([0 7]);
    elseif selectact == 4;
        selectobs = randi([1 6]);
    end
    teaminst(m:m+selectobs, 1) = selectact;
    m = m+selectobs;
  end
  storeteams(:, i) = teaminst(1:60, 1);
  i = i + 1;
  clear teaminst m selectact selectobs
  end
dlmwrite('simulated string data.txt', storeteams);
```

To conduct meaningful validation, you must decide which parameters need to reflect the real world and how you will test them. Specific validation parameter selection depends on the type of study being conducted, but typically includes the variables that will be manipulated during virtual experimentation and the results. These parameters will be compared to the empirical data using common statistical comparison tests (e.g., χ^2, t-test) or other non-parametric tests (e.g., Wilcoxon rank sum test, Mann-Whitney test). The specific tests will be dependent upon the distribution of the data being tested.

For our basic example with continuous data, we are interested in the relationship between cohesion and team effectiveness. Re-running the regression analyses on the simulated data and comparing the results to the regression results obtained using the original empirical data is a viable approach here for validating the simulation procedures. If the regression coefficients from the simulated data have similar signs as the coefficients from the empirical data and the magnitude of the coefficients fall within a 95 % confidence interval, then we may proceed to virtual experimentation. The level of confidence should be chosen based on what is appropriate for your study. Table 8.1 shows the actionable code for validating the simulation procedures in this way.

For the discrete longitudinal data simulation basic example, multiple parameters will need to be validated including the point when cohesion emerges, the number of minutes devoted to cohesion, and the relationship between cohesion and effectiveness. Table 8.4 provides the actionable code for validating the discrete data simulation procedures on these characteristics by comparing empirical data from ten teams to similar data from the simulated teams. The specific comparisons are when cohesion first surfaces, the number of cohesion observations per communication string, and performance. Performance is represented as a curvilinear relationship with cohesion. We generated a regression model of this cohesion-performance relationship using the empirical data. This model is used to calculate the performance of simulated teams by entering the point where cohesion surfaces into the regression model. These three characteristics of the simulated communication strings are then compared to the empirical data using the Wilcoxon rank sum test and the results are output. While the validation should indicate that the simulated procedures produce communication strings that are representative of the real world, you may also decide that you need to validate other parameters. For example, if you think you may be interested in inspecting the ways other topics are affected when cohesion is manipulated, then you will need to validate the occurrences of all the possible activity states, rather than only focusing on cohesion.

As we suggested in Fig. 8.1, if the evidence supports validating the simulation procedures, then the next step is virtual experimentation. Unfortunately, validation is not always easily achieved, particularly on the first try. If your testing does not support the validation, then you will want to stop and evaluate why the validation was not successful. We suggest that you have two options: (1) fix your simulation procedures or (2) go back to the beginning and reformulate your what-if research question and/or secure more (or different) empirical data. Common problems come from issues with the initial assumptions (e.g., normality, conditional relationships, linear relationships), miscalculations using the empirical data, or coding errors. You should start by checking for coding errors or miscalculations of parameters that can easily be fixed with corrective action.

Table 8.4 Discrete data simulation: validation

```
%Validation step
dlmread('simulated string data.txt');
firstcohesion = zeros(10000, 1);
freqcohesion = zeros(10000, 1);
for i = 1:10000;
    k = find(storeteams(:, i)==2);
    [row, col] = size(k);
    if row == 0;
    firstcohesion(i, 1) = 0;
    freqcohesion(i, 1) = 0;
    else
    firstcohesion(i, 1) = k(1, 1);
    freqcohesion(i, 1) = row;
    end
    clear k row;
end
%Test first cohesion
Empfirst = [1, 1, 2, 2, 4, 8, 10, 12, 30, 45];
[pfirst,h,stats] = ranksum(Empfirst, firstcohesion);
if pfirst >= 0.05
    p = num2str(pfirst, 3);
    fprintf('Cohesion first point validation a success, p = %s \n', p)
else
    display('Cohesion first point validation error, check and retry')
end
%Test frequency of cohesion
Empfreq = [5, 8, 9, 10, 10, 11, 11, 12, 15, 20];
[pfreq,h,stats] = ranksum(Empfreq,freqcohesion);
if pfreq >= 0.05
    p = num2str(pfreq, 3);
    fprintf('Cohesion frequency validation a success, p = %s \n', p)
else
    display('Cohesion frequency validation error, check and retry')
end
%Relationship of cohesion-performance
Empperf = [2002.5 1992 1986.5 1980 1980 1972.5 1972.5 1964 1932.5 1860];
Simperf = -0.5 * firstcohesion.^2 + 3 * firstcohesion + 2000;
[pperf,h,stats] = ranksum(Empperf,Simperf);
if pperf >= 0.05
    p = num2str(pperf, 3);
    fprintf('Cohesion-performance relationship validation a success, p = %s
\n', p)
else
    display('Cohesion-performance relationship validation error, check and
retry')
end
```

You may decide that more simulation procedures are necessary. When you have a complex phenomenon with a lot of moving parts, you may start with the simplest set of simulation procedures for the sake of parsimony. But, through validation testing you may determine that more procedures are needed. For example, in the second basic example where we generate activity strings, perhaps a division at the midpoint does not provide enough fidelity, and the distributions actually change across quartiles. Alternatively, the activities may have more interrelationships than anticipated and conditional probabilities may be more appropriate (e.g., planning is generally followed by conflict). Whatever the reason, sometimes developing a valid set of simulation procedures requires additional information and/or steps to ensure statistical similarity to the real-world data.

More insidious are the irregularities in the empirical data that can become amplified in the simulation when the generated sample is large. These issues can challenge your assumptions about how to characterize variables, the covariances between variables, and any statistical relationships among the variables. Such issues may be a sign that perhaps you need to take a step back to re-think theoretical connections, re-characterize the originally tested relationship, reconsider the what-if research question being studied, or realize that the original data are not representative of the real-world.

Step 4. Conduct virtual experiments

Purpose: Compare simulation results across varying conditions to inform theory construction and future research activities

Actions: Manipulate simulation inputs to ascertain how results change under varying conditions

Outputs: Results that answer the original what-if research questions and provide insights into future theory and research development

Once you have validated your simulation procedures you can conduct virtual experimentation. Specifically, you can now generate datasets with manipulated conditions by changing parameters in your simulation procedures. The way the parameters are changed should be based on your theoretical understanding of the phenomena and your what-if research questions. To determine the impact of the manipulations, the simulated datasets can be compared to one another and to a baseline. Often times, the empirical data is used to represent the baseline case.

For the basic example using continuous data from a survey, the virtual experiments will focus on the cohesion-team effectiveness relationship under different synchronicity scenarios. We partition the empirical data into teams that communicated face-to-face a majority of the time (i.e., a high synchronicity condition) versus teams that engaged in computer-mediated communication a majority of the time (i.e., a low synchronicity condition). From the partitioned dataset we can calculate the mean and covariance information for these specific conditions; for example, we assume that the means are as follows for high synchronicity $\mu_{hs} = (4.5, 8)$ and low synchronicity $\mu_{ls} = (2, 5)$, and the covariance matrices are:

$$\sum_{hs} = \begin{bmatrix} 2.4 & 2.5 \\ 2.5 & 4.2 \end{bmatrix}, \quad \sum_{ls} = \begin{bmatrix} 2.1 & 1.7 \\ 1.7 & 3.2 \end{bmatrix}$$

Table 8.1 shows the implementation of virtual experimentation in the actionable computer code with the new parameters from the partitioned dataset. Using regression analysis on the simulated data, we obtain results that can be compared to the baseline condition. For example, in Fig. 8.3, plots of regression results across the different conditions are shown. This type of plot can be used to ascertain the influence of synchronicity on the cohesion-effectiveness relationship.

For the basic example using discrete data from a survey, the virtual experiments will focus on the cohesion-team effectiveness relationship when cohesion occurs earlier or later. We manipulate the probability of selecting cohesion in minutes 1–30 by considering two interventions. Specifically, we examine the case that man-

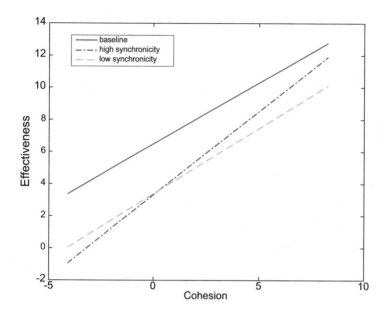

Fig. 8.3 Continuous data simulation: virtual experimentation results

agers prompt cohesion building discussions instead of planning, or vice versa. In Table 8.5, the actionable code for trying these two cohesion scenarios is presented; the first increases the probability of selecting cohesion while decreasing the probability of selecting planning; and the second scenario decreases the probability of selecting cohesion while increasing the probability of selecting planning. To test the effects on the cohesion-performance relationship, the simulated communication strings under each scenario are evaluated for when cohesion surfaces. These data are then evaluated using the cohesion-performance model. Finally, a comparison among scenarios and the empirical data indicate that significant differences exist, at least between the second scenario and other datasets. Figure 8.4 shows an example of the boxplot and ANOVA findings from the comparisons. These results may help provide the theoretical impetus to pursue research about prompting cohesion activities and/or delaying planning activities.

8.3   Example Applications

We now turn to two more complex examples to demonstrate how simulation and virtual experimentation can be done using empirical data. The two basic examples we worked through in the tutorials are simplified versions of these two published simulation studies. In Kennedy et al. (2011), we simulate continuous data using an empirical dataset that was generated from employee questionnaires. We study the curvilinear relationship between team communication and performance under various levels of project complexity. In Kennedy and McComb (2014), we simulate

Table 8.5 Discrete data simulation: virtual experimentation

```
%Virtual Experimentation - Change the probability of selecting cohesion in
first 30 minutes%
VEstoreteams = zeros(1, 10000);
selectobs = 0;
selectact = 0;
scenario = [0.575, 0.7000; 0.175, 0.300];
for s = 1:2;
  storeteams = zeros(60, 10000);
  for i = 1:10000;
  teaminst = []; m = 1;
  %for first 30 minutes%
  while m < 30;
    %Selects Activity 1, 2, 3, 4%
    selectless30 = rand();
    if selectless30 <= 0.125
        selectact = 1;
    elseif selectless30 <= scenario(s, 1);
        selectact = 2;
    elseif selectless30 <= scenario(s, 2);
        selectact = 3;
    elseif selectless30 <= 1.000;
        selectact = 4;
    end
    %Selects Activity observation length%
    if selectact == 1
        selectobs = randi([1 4]);
    elseif selectact == 2;
        selectobs = randi([2 4]);
    elseif selectact == 3;
        selectobs = randi([2 8]);
    elseif selectact == 4;
        selectobs = randi([0 8]);
    end
    %Fills in the team instance string%
    teaminst(m:m+selectobs, 1) = selectact;
    m = m+selectobs;
  end
  %for second 30 minutes%
  while m < 60
  %Selects Activity 1, 2, 3, 4%
  selectact = randi([1 4]);
    %Selects Activity observation length%
    if selectact == 1;
        selectobs = randi([2 6]);
    elseif selectact == 2;
        selectobs = randi([1 4]);
    elseif selectact == 3;
        selectobs = randi([0 7]);
    elseif selectact == 4;
        selectobs = randi([1 6]);
    end
    teaminst(m:m+selectobs, 1) = selectact;
    m = m+selectobs;
  end
  storeteams(:, i) = teaminst(1:60, 1);
  i = i + 1;
```

(continued)

Table 8.5 (continued)

```
    clear teaminst m selectact selectobs
    end
    VEstoreteams = (cat(1, VEstoreteams, storeteams));
    clear storeteams
end
VEstoreteams = VEstoreteams(2:121, :);
dlmwrite('virtualexperiment string data.txt', VEstoreteams);

%Relationship of cohesion-performance
VEscenario1 = VEstoreteams(1:60, :);
VEscenario2 = VEstoreteams(61:120, :);
for i = 1:10000;
    k = find(VEscenario1(:, i)==2);
    [row, col] = size(k);
    if row == 0;
    VE1firstcohesion(i, 1) = 0;
    else
    VE1firstcohesion(i, 1) = k(1, 1);
    end
    clear k row;
end
for i = 1:10000;
    k = find(VEscenario2(:, i)==2);
    [row, col] = size(k);
    if row == 0;
    VE2firstcohesion(i, 1) = 0;
    else
    VE2firstcohesion(i, 1) = k(1, 1);
    end
    clear k row;
end

Empperf = [2002.5 1992 1986.5 1980 1980 1972.5 1972.5 1964 1932.5 1860];
VE1first = sort(VE1firstcohesion)'; VE2first = sort(VE2firstcohesion)';
VE1perf = -0.5 * VE1first.^2 + 3 * VE1first + 2000;
VE2perf = -0.5 * VE2first.^2 + 3 * VE2first + 2000;

display('Comparing Performance Results across Baseline and Virtual
Experiments')
groups = [ones(1, 10000) ones(1, 10000)*2 ones(1, 10)*3];
perf = [VE1perf VE2perf Empperf];
[p,t,st] = anova1(perf, groups);
[c,m,h,nms] = multcompare(st, 'ctype', 'hsd')
```

discrete data in the form of communication strings using an empirical dataset that was generated by coding transcribed team communication from observed laboratory teams. We conduct virtual experiments to determine how the timing of when teams shift among processes impacts team performance.

8.3.1 Team Communication, Performance, and Project Complexity

Step 1: Formulate what-if research questions and secure empirical data. The impetus for this study came by combining the ideas and results from two studies conducted using cross-sectional survey data. So, in this case, we had empirical data that

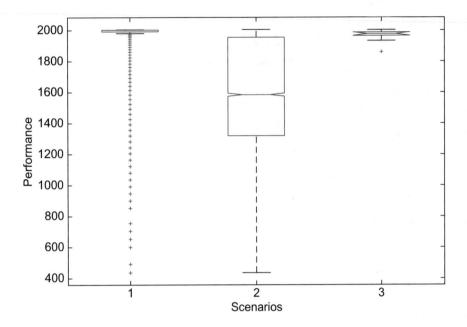

Fig. 8.4 Discrete data simulation: virtual experimentation results

led us to a what-if research question. In the first study, Patrashkova-Vozdolska et al. (2003) found a curvilinear relationship between team communication and performance that was different depending on the media used by the team to communicate. In the second, McComb, Green, and Compton (2007) characterized project complexity as having two dimensions (i.e., multiplicity (having multiple options for accomplishing the project and/or possible end states to satisfy) and ambiguity (i.e., conflict and/or uncertainty associated with options and end states)) and found that project complexity moderated the relationship between team flexibility and performance. We were then interested in delving further into the team communication-performance relationship and wondered how project complexity might impact it. We did not, however, have an adequate sample to test this relationship using the empirical data available. We decided to develop simulation procedures and conduct virtual experiments to test the what-if research question: *what if project complexity levels were different, how might the communication-performance relationship be affected?*

To test this what-if research question, we decided to simulate team-level data about how frequently various communication media are used under different levels

Table 8.6 Variables in team communication, performance, and project complexity example

Variable	Definition	Measurement	Sample means for high ambiguity high multiplicity condition
Team performance			
Goal achievement	Meet technical objectives and business goals	1=low to 5=high	3.69
Efficiency	Meet cost/time estimates	1=low to 5=high	2.98
Communication media			
Email		Frequency of use on a scale from 1=never to 5=often	3.95
Telephone			4.20
Face-to-face			4.67
Control variables			
Task significance	Project importance to themselves and the organization	1=low to 5=high	Control variable means are not reported in the paper
Team size	Number of fulltime members		
Co-location	Three levels: same site; same city, different sites; different cities, states, or countries	Two dummy coded variables	
Project complexity			
Multiplicity	Having multiple options for accomplishing the project and/or possible end states to satisfy	Median splits used to partition data into high/low complexity	
Ambiguity	Conflict and/or uncertainty associated with options and end states		

of project complexity. This data could then be used to replicate the multivariate regression analysis from Patrashkova-Volzdolska et al. (2003), where team performance was regressed on four control variables, three linear terms representing communication media (i.e., email, telephone, and face-to-face), and a squared term for each communication media. Regression analysis was conducted for each level of project complexity.

The empirical dataset available was from a field study and contained team-level data from 60 cross-functional teams. The specific variables simulated in this study are listed in Table 8.6. These variables represent the variables in the regression equations reported in Patrashkova-Volzdolska et al. (2003) and the project complexity variables reported in McComb et al. (2007). Mean vectors were computed for the performance, communication, and control variables. These vectors were then used to compute partitioned covariance matrices, where performance variables were treated as multivariate normal conditional on the media and control variables. Kennedy et al. (2011) provides a thorough explanation of how the covariance matrices were constructed.

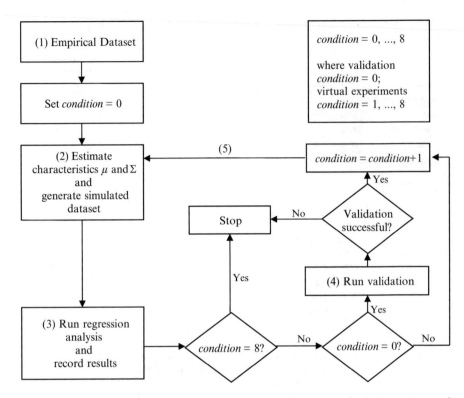

Fig. 8.5 Flowchart of simulation procedures. Reprinted from *Journal of Engineering and Technology Management*, 28, Kennedy, DM, McComb, SA, & Vozdolska RR, An investigation of project complexity's influence on team communication using Monte Carlo simulation, 109–127, 2011, with permission from Elsevier

Step 2: Develop simulation procedures. We applied the Monte Carlo method for multivariate normal sampling using mean (μ) and covariance (Σ). Our sample size was 10,000 simulation runs of 100 simulated teams each. This sample size was based on the guidance of Gorsuch (1983), who recommends approximately ten observations per independent variable for regression analysis, and Law and Kelton (2000) who recommend 10,000 simulation runs. A flowchart of the simulation procedures can be seen in Fig. 8.5. These procedures were coded in R (version 2.5).

Step 3: Validate simulation procedures. To validate our simulation procedures, we computed the mean vector and partitioned covariance matrix using the entire empirical dataset (i.e., no partitions for project complexity). We executed 10,000 simulation runs containing 100 simulated teams each. In other words, we simulated 10,000 samples with n=100 teams in each sample. For each sample, we estimated regression parameters and averaged each regression estimate, *p*-value, and R^2 across the 10,000 samples. The results of the simulated data patterned the empirical data. For example, when efficacy was the performance variable, both email and email^2 were significant and the direction of the coefficients were the same (positive for email and negative for email^2).

Step 4: Conduct virtual experiments. To answer our what-if research question about the impact of project complexity on the communication-performance relationship, the empirical dataset was partitioned based on the project complexity reported by the teams. Median values of multiplicity and ambiguity were used to determine if a team in the empirical dataset was completing a project of high or low complexity. Eight experimental conditions were examined: high multiplicity-high ambiguity, high multiplicity-low ambiguity, low multiplicity-high ambiguity, low multiplicity-low ambiguity, high multiplicity only, low multiplicity only, high ambiguity only, and low ambiguity only. For each condition, 10,000 samples of 100 teams each were generated and the regression results were computed using the same approach as we used for validation. The results suggest that in many cases the same inverted curvilinear relationship between communication and performance can be expected. In some cases, however, more communication may be better. For example, teams may be more efficient if they communicate face-to-face when ambiguity is very low or very high.

8.3.2   Team Performance and Process Shifts

Step 1: Formulate what-if research questions and secure empirical data. In this example, we had a what-if research question that led us to generate an empirical dataset. While conducting laboratory simulations where teams of three undergraduate students completed a scheduling task, we began discussing how processes unfold over time. We knew of the research suggesting that directing team communication toward certain topics (e.g., Okhuysen, 2001; Okhuysen & Eisenhardt, 2002; Okhuysen & Waller, 2002) at specific points in time (e.g., Gersick, 1988, 1989; Okhuysen & Waller, 2002; Katzenbach & Smith, 1999) may enhance team performance. At the same time, we were familiar with Marks, Mathieu, and Zaccaro (2001) recurring phase model of team activities, where teams work through transition and action phases. Pulling all of this evidence together, we decided to pose the what-if research question: *what if interventions affected team communication about processes, would it change (i.e., help/hinder) the relationship between when process shifts occur and team performance?*

Answering this what-if research question required figuring out what processes to manipulate, when to manipulate them, and sample sizes needed to meaningfully test the various conditions. We decided to turn to simulation and virtual experimentation to guide these decisions. Table 8.7 describes the variables of interest in our study. By using the videotapes of the 60 teams completing our laboratory simulations for a different research purpose, we were able to generate an empirical dataset by transcribing and coding the team conversations, where each message exchanged among team members contained information about one topic. The communication strings were a series of numbers (i.e., 10, 20, 30, 40, 50), where each number represented a specific topic discussed by the teams. In other words, we had a set of discrete data points representing the communication of each team. The resulting empirical dataset

Table 8.7 Variables in team performance and process shifts example

Variables	Definition	Measurement	Virtual experiments
Processes			
Mission analysis	Task objectives, context, resources	Pattern recognition of shift point (i.e., message number)	Increase probability of selection in quartile one
Goal specification	Task goals and associated priorities		n/a
Tactical strategy	Approaches that could be used to complete task		Increase probability of selection in quartile one
Operational strategy	Division of labor among team members		n/a
Action process	Activities that move task toward completion		Delay probability of selection to quartile two or three
Performance			
Schedule cost	Cost of schedule developed by team	Dollars	Calculate using a neural network
Time	Time to complete schedule	Minutes	

contained communication strings and the corresponding team-level cost and time performance data. The cost performance was calculated from the workforce schedule created by laboratory teams. The time performance captures the number of minutes the laboratory team took to complete the scheduling task.

Step 2: Develop simulation procedures. Our initial investigation of the empirical data suggested that topics were discussed in different proportions across quartiles of communication. To ensure we were able to generate realistic communication strings, we calculated several distributions representing a variety of characteristics of the communication strings, including communication string length, topics per quartile, and the number of observations per topic per quartile. The simulation procedures included eight steps: (1) select the length of these communication string; (2) divide the string length so that the program can trace quartile completion and change to the appropriate distributions as the communication string progresses through quartiles; (3) go to the first observation point of the communication string; (4) call up the distributions about topic; (5) select the topic to be inserted into the communication string; (6) select the number of observations to be assigned to the topic; (7) insert the selected topic for the selected number of observations; and (8) update the position point in the communication string and check whether the quartile length and/or string length has been reached. Once the fourth quartile length was complete, and thereby the communication string length was achieved, the communication string was stored in a dataset, the variables cleared, and the simulation readied to create another string. The simulation completed when 10,000 communication strings had been generated per the suggestion by Law and Kelton (2000). The process flowchart can be seen in Kennedy and McComb (2014). The simulation procedures were programmed in Matlab.

Step 3: Validate simulation procedures. To validate the procedures, we executed our simulation procedures to generate communication strings for 10,000 simulated teams. We then evaluated the simulated communication strings on six characteristics: (1) communication string length, (2) topics selected per quartile, (3) number of observations of a topic per quartile, (4) the frequency of transitions from one topic to the next topic selected by quartile, (5) when teams first shifted topics, and (6) the performance associated with communication strings. The first four evaluations were straight-forward counts from each communication string. Evaluating when teams shifted topics was more complicated. We considered teams to have conducted a process shift when at least three messages were exchanged about a topic followed by at least 25 messages about other topics, because the process shift indicates "the point in time when teams complete their focus on a process and change to focus on one or more different processes" (Kennedy & McComb, 2014, p. 784). Each communication string was evaluated using a pattern recognition sub-routine in order to identify when the shift point occurred for each topic.

The final evaluation required generating performance scores for each simulated communication string using the points where process shifts occurred. In preparation for this step, we trained, tested, and validated a neural network model using Matlab's Neural Network Toolbox. The model links the points in the team communication string where process shifts occur to cost and time performance using the empirical data from the laboratory teams. A complete description of the model can be found in Kennedy and McComb (2014). The model was used to estimate the cost and time performance output for simulated teams based on the process shift points in the simulated communication strings.

Once we obtained information about the communication string characteristics, we compared the distributions from communication strings of the simulated teams to those of the laboratory teams. Where distributions were not normally distributed, we used the Wilcoxon rank sum statistical test to compare the data between simulated teams and those of the laboratory teams, otherwise a t-test was applied. We found no significant difference between simulated communication string characteristics and those of laboratory teams in terms of communication string length, topics selected per quartile, number of observations of a topic per quartile, or when teams first shifted topics and the performance associated with communication strings (i.e., all comparisons with $p > 0.05$). Of the 100 comparisons of how frequently teams transitioned from one topic to the next (i.e., 5 topics connecting to 5 topics including itself across 4 quartiles), 77 were not significantly different. Even with the disappointing results for transition frequency, we concluded that our simulation procedures were adequate in generating communications strings that were representative of the real-world teams.

Step 4: Conduct virtual experiments. To answer our what-if research question about the impact of process shifts on team performance, we manipulated communication by changing the probabilities depicting when a topic might be discussed.

Specifically, we tested eight different experimental conditions that are grounded in our theoretical understanding of what communication patterns may enhance team performance: (1) initialize mission analysis by coding it as the first message exchanged by teams, (2) initialize strategy formulation, (3) delay action processes by one quartile by setting the probability of exchanging an action message during the first quartile to 0; (4) delay action processes by two quartiles; (5, 6, 7, & 8) initialize mission analysis or strategy formulation *and* delay action by one or two quartiles. For each experimental condition, 10,000 simulated teams were generated. Each simulated team's scaled performance measure was used to test for differences across the conditions. The results suggest that delaying action until after the midpoint of the team's life cycle may be the most effective intervention.

8.4 Conclusion

Simulation and virtual experimentation are accessible tools in a researcher's arsenal, as we have demonstrated in this chapter. By systematically working through the steps we presented in our tutorial and demonstrate in the Kennedy et al. (2011) and Kennedy and McComb (2014) journal articles, researchers can construct simulation procedures grounded with continuous or discrete empirical data to answer what-if questions. Simulation by no means replaces real-world investigations. But when valid simulation procedures are used to conduct virtual experiments, the results have the potential to help researchers construct theory and design future empirical work.

References

Anderson Jr., E. G., & Lewis, K. (2013). A dynamic model of individual and collective learning amid disruption. *Organization Science, 25*(2), 356–376.
Avramidis, A. N., & L'Ecuyer, P. (2006). Efficient Monte Carlo and quasi—Monte Carlo option pricing under the variance gamma model. *Management Science, 52*(12), 1930–1944.
Bhuiyan, N., Gerwin, D., & Thomson, V. (2004). Simulation of the new product development process for performance improvement. *Management Science, 50*(12), 1690–1703.
Burton, R. M., & Obel, B. (2011). Computational modeling for what-is, what-might-be, and what-should-be studies-And triangulation. *Organization Science, 22*(5), 1195–1202.
Carley, K., & Prietula, M. (Eds.) (1994). *Computational organization theory.* Hillsdale, NJ: Erlbaum.
Chance, D. M., Hillebrand, E., & Hillard, J. E. (2008). Pricing an option on revenue from an innovation: An application to movie box office revenue. *Management Science, 54*(5), 1015–1028.
Charnes, J. M., & Shenoy, P. P. (2004). Multistage Monte Carlo method for solving influence diagrams using local computation. *Management Science, 50*(3), 405–418.
Davis, J. P., Eisenhardt, K. M., & Bingham, C. B. (2007). Developing theory through simulation methods. *The Academy of Management Review, 32*, 480–499.
Dennis, A. R., Fuller, R. M., & Valacich, J. S. (2008). Media, tasks, and communication processes: A theory of media synchronicity. *MIS Quarterly, 32*(3), 575–600.

Georgieva, A., & Jordanov, I. (2009). Global optimization based on novel heuristics, low-discrepancy sequences and genetic algorithms. *European Journal of Operational Research, 196*, 413–422.

Gersick, C. J. G. (1988). Time and transition in work teams: Toward a new model of group development. *Academy of Management Journal, 31*, 9–41.

Gersick, C. J. G. (1989). Marking time: Predictable transitions in task groups. *Academy of Management Journal, 32*, 274–309.

Gorsuch, R. L. (1983). *Factor Analysis* (2nd ed.,). Hillsdale, NJ: Lawrence Erlbaum.

Grand, J. A., Braun, M. T., Kuljanin, G., Kozlowski, S. W. J., & Chao, G. T. (2016). The dynamics of team cognition: A process-oriented theory of knowledge emergence in teams. *Journal of Applied Psychology, 101*(10), 1353–1385.

Harrison, J. R., Lin, Z., Carroll, G. R., & Carley, K. M. (2007). Simulation modeling in organizational and management research. *The Academy of Management Review, 32*, 1229–1245.

Hulin, C. L., & Ilgen, D. R. (2000). Introduction to computational modeling in organizations: The good that modeling does. In D. R. Ilgen, & C. L. Hulin (Eds.), *Computational modeling of behavior in organizations: The third scientific discipline* (pp. 3–18). Washington, DC: American Psychological Association.

Katzenbach, J., & Smith, D. (1999). *The wisdom of teams: Creating the high-performance organization*. Cambridge, MA: Harvard Business School Press.

Kennedy, D. M., & McComb, S. A. (2014). When teams shift among processes: Insights from simulation and optimization. *Journal of Applied Psychology, 99*(5), 784–815.

Kennedy, D. M., McComb, S. A., & Vozdolska, R. (2011). Using simulation to analyze complex behavioral models: An investigation of project complexity's influence on team communication. *Journal of Engineering and Technology Management., 28*(3), 109–127.

Kozlowski, S. W., & Chao, G. T. (2012). The dynamics of emergence: Cognition and cohesion in work teams. *Managerial and Decision Economics, 33*(5–6), 335–354.

Kozlowski, S. W., Chao, G. T., Grand, J. A., Braun, M. T., & Kuljanin, G. (2013). Advancing multilevel research design capturing the dynamics of emergence. *Organizational Research Methods, 16*(4), 581–615.

Kozlowski, S. W. J., Chao, G. T., Grand, J. A., & Braun, M. T. (2016). Capturing the multilevel dynamics of emergence: Computational modeling, simulation, and virtual experimentation. *Organizational Psychology Review, 6*(1), 3–33.

Larson Jr., J. R. (2012). Computer simulation methods for groups. In A. B. Hollingshead, & M. S. Poole (Eds.), *Research methods for studying groups and teams* (pp. 329–357). New York, NY: Routledge.

Law, A. M., & Kelton, W. D. (2000). *Simulation modeling and analysis* (3rd ed.,). New York, NY: McGraw-Hill.

Marks, M. A., Mathieu, J. E., & Zaccaro, S. J. (2001). A temporally based framework and taxonomy of team processes. *Academy of Management Review, 26*, 356–376.

McGrath, J. E. (1981). Dilemmatics: The study of research choices and dilemmas. *The American Behavioral Scientist, 25*(2), 179.

McComb, S. A., Green, S. G., & Compton, W. D. (2007). Team flexibility's relationship to staffing and performance in complex projects: An empirical analysis. *Journal of Engineering Technology Management, 24*, 293–313.

Metropolis, N., & Ulam, S. (1949). The monte carlo method. *Journal of the American Statistical Association, 44*(247), 335–341.

Newman, D. A. (2003). Longitudinal modeling with randomly and systematically missing data: A simulation of ad hoc, maximum likelihood, and multiple imputation techniques. *Organizational Research Methods, 6*(3), 328–362.

Okhuysen, G. A. (2001). Structuring change: Familiarity and formal interventions in problem-solving groups. *Academy of Management Journal, 44*, 794–808.

Okhuysen, G. A., & Eisenhardt, K. E. (2002). Integrating knowledge in groups: How formal interventions enable flexibility. *Organization Science, 13*, 370–386.

Okhuysen, G. A., & Waller, M. J. (2002). Focusing on midpoint transitions: An analysis of boundary conditions. *Academy of Management Journal, 45,* 1056–1065.

Patrashkova, R., & McComb, S. A. (2004). Exploring why more communication is not better: Insights from a computational model of cross-functional teams. *Journal of Engineering and Technology Management, 21*(1–2), 23–81.

Patrashkova-Volzdoska, R. R., McComb, S. A., Green, S. G., & Compton, W. D. (2003). Examining a curvilinear relationship between communication frequency and team performance in cross-functional project teams. *IEEE Transactions on Engineering Management, 50*(3), 262–269.

Prietula, M., Carley, K., & Gasser, L. (Eds.) (1998). *Simulating organizations: Computational models of institutions and groups.* Cambridge, MA: MIT Press.

Ren, Y., Carley, K. M., & Argote, L. (2006). The contingent effects of transactive memory: When is it more beneficial to know what others know? *Management Science, 52*(5), 671–682.

Roth, P. L., Switzer III, F. S., & Switzer, D. M. (1999). Missing data in multiple item scales: A Monte Carlo analysis of missing data techniques. *Organizational Research Methods, 2,* 211–232.

Solow, D., Vairaktarakis, G., Piderit, S. K., & Tsai, M. (2002). Managerial insights into the effects of interactions on replacing members of a team. *Management Science, 48*(8), 1060–1073.

Valenzuela, J. M., & Mazumbar, M. (2003). Commitment of electric power generators under stochastic market prices. *Operations Research, 51*(6), 880–893.

Wang, M., Zhou, L., & Zhang, Z. (2016). Dynamic modeling. *Annual Review of Organizational Psychology and Organizational Behavior, 3,* 241–266.

Printed in the United States
By Bookmasters